Geometry in Figures

Arseniy Akopyan

Arseniy Akopyan. Geometry in Figures.

This book is a collection of theorems and problems in classical Euclidean geometry formulated in figures. It is intended for advanced high school and undergraduate students, teachers and all who like classical geometry.

Cover art created by Maria Zhilkina.

Copyright ©2011 Arseniy Akopyan
All rights reserved.
ISBN-13: 978-1463570743
ISBN: 1463570740

Contents

1 Elementary theorems 6
2 Triangle centers 9
3 Triangle lines 15
4 Elements of a triangle 18
   4.1 Altitudes of a triangle 18
   4.2 Orthocenter of a triangle 21
   4.3 Angle bisectors of a triangle 23
   4.4 The symmedian and its properties 27
   4.5 Inscribed circles 29
   4.6 Inscribed and circumscribed circles of a triangle 38
   4.7 Circles tangent to the circumcircle of a triangle 39
   4.8 Circles related to a triangle 42
   4.9 Concurrent lines of a triangle 50
   4.10 Right triangles 55
   4.11 Theorems about certain angles 55
   4.12 Other problems and theorems 57
5 Quadrilaterals 59
   5.1 Parallelograms 59
   5.2 Trapezoids 61
   5.3 Squares 62
   5.4 Circumscribed quadrilaterals 63
   5.5 Inscribed quadrilaterals 66
   5.6 Four points on a circle 68
   5.7 Altitudes in quadrilaterals 71
6 Circles 73
   6.1 Tangent circles 73
   6.2 Monge's theorem and related constructions 75
   6.3 Common tangents of three circles 79
   6.4 Butterfly theorem 81
   6.5 Power of a point and related questions 82
   6.6 Equal circles 84
   6.7 Diameter of a circle 84
   6.8 Constructions from circles 86
   6.9 Circles tangent to lines 89
   6.10 Miscellaneous problems 90
7 Projective theorems 94
8 Regular polygons 96

	8.1 Remarkable properties of the equilateral triangle	98
9	Appended polygons	101
10	Chain theorems	104
11	Remarkable properties of conics	109
	11.1 Projective properties of conics	112
	11.2 Conics intersecting a triangle	117
	11.3 Remarkable properties of the parabola	118
	11.4 Remarkable properties of the rectangular hyperbola	120
12	Remarkable curves	121
13	Comments	123

Preface

This book is a collection of theorems (or rather facts) of classical Euclidean geometry formulated in figures.

The figures were drawn in such a way that the corresponding statements can be understood without any additional text.

We usually draw primary lines of each problem using bold lines. Final conclusions are illustrated by dashed lines. Centers of circles, polygons and foci of conics are denoted by points with a hole. Bold lines in the section about conics denote directrices of conics.

It is commonly very hard to determine who the author of certain results is. There are comments at the end of the book, most of which refer to the source of a result. Some of the results were discovered by the author while working on this book, but these results are probably not new.

I thank Dmitry Shvetsov, Fedor Petrov, Pavel Kozhevnikov and Ilya Bogdanov for useful remarks. I'm grateful to Mikhail Vyalyi for helping me to learn the METAPOST system, which was used to draw all the figures, except the cover which has been made by the admirable Maria Zhilkina.

If you have any comments or remarks or you spot any mistakes, please send an email to *arseny.akopyan@gmail.com*.

1 Elementary theorems

1.1)

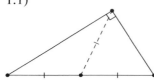

1.2) **Pythagorean theorem**

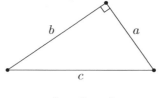

$$a^2 + b^2 = c^2$$

The inscribed angle theorem

1.3)

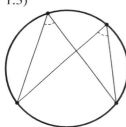

1.4)

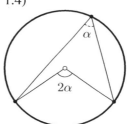

1.5)

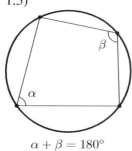

$$\alpha + \beta = 180°$$

1.6)

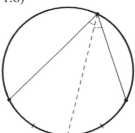

1.7)

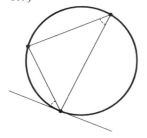

1.8) **Miquel's theorem**

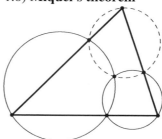

1.9)
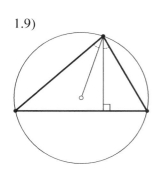

1.10)
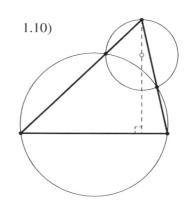

1.11)
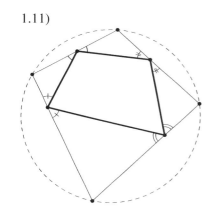

1.12)
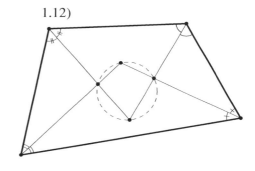

1.13)
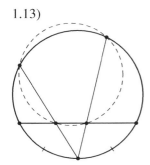

1.14)
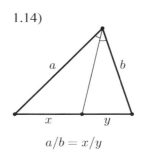

$a/b = x/y$

1.15)
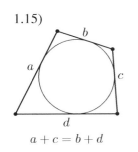

$a + c = b + d$

1.16)
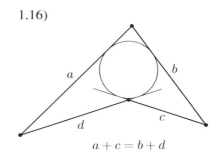

$a + c = b + d$

1.17)
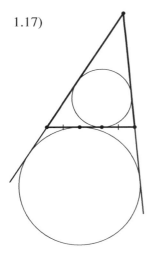

1.18)
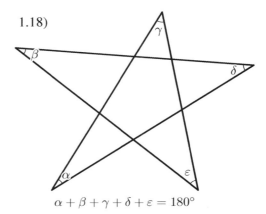
$\alpha + \beta + \gamma + \delta + \varepsilon = 180°$

1.19)
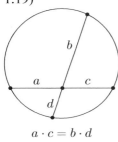
$a \cdot c = b \cdot d$

1.20)
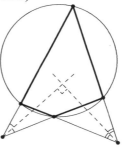

1.21)
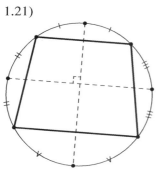

2 Triangle centers

2.1)

2.2)

2.3)

2.4) **Gergonne point**

2.5)

2.6) **Lemoine point**

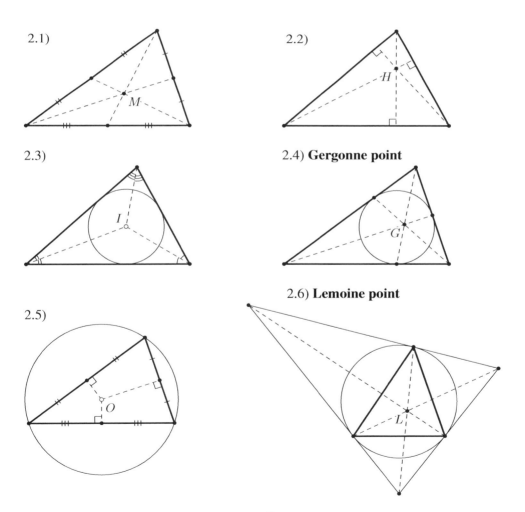

2.7) Nagel point

2.8)

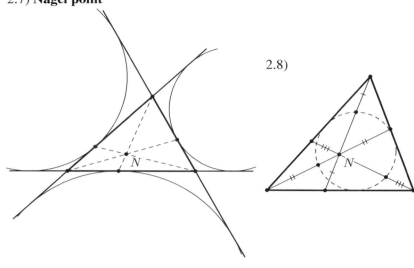

2.9) First Torricelli point

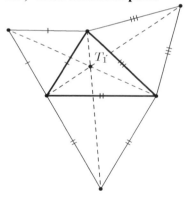

2.10) Second Torricelli point

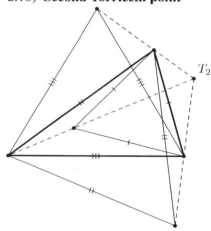

2.11) First Apollonius point

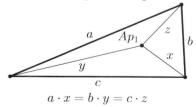

$a \cdot x = b \cdot y = c \cdot z$

2.12) Second Apollonius point

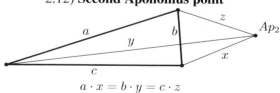

$a \cdot x = b \cdot y = c \cdot z$

2.13) First Soddy point

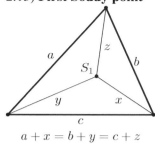

$a + x = b + y = c + z$

2.14) Second Soddy point

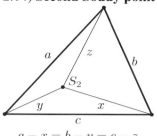

$a - x = b - y = c - z$

2.15)

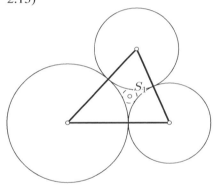

2.16)

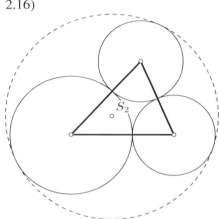

2.17)

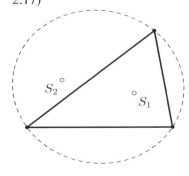

2.18)

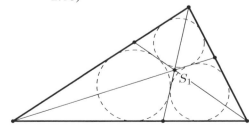

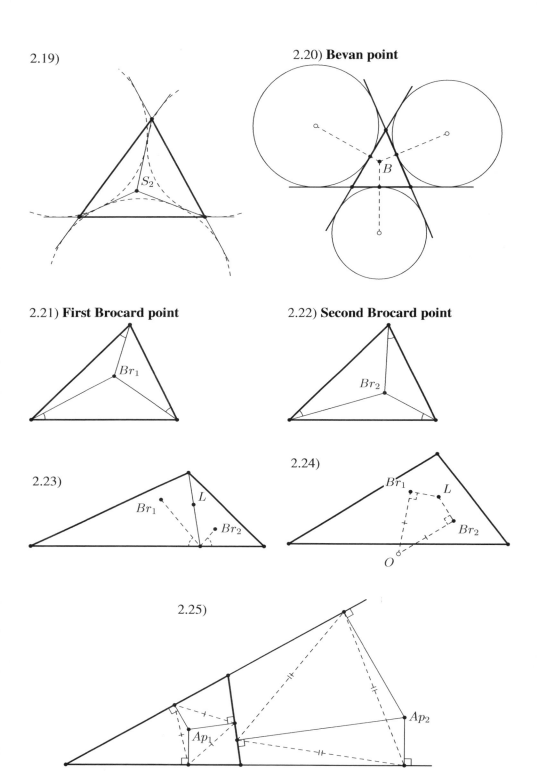

2.26)
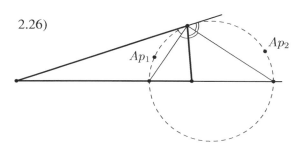

2.27)
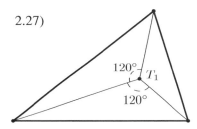

2.28)
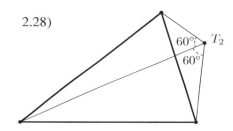

2.29)
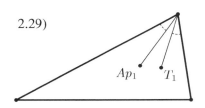

2.30)
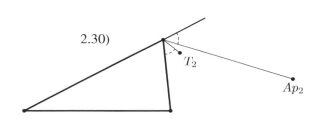

2.31) **First Lemoine circle**
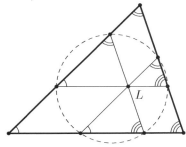

2.32) **Second Lemoine circle**
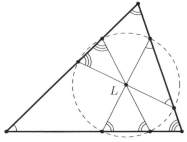

Miquel point and its properties

2.33) **Miquel point** 2.34)

2.35) **Clifford's circle theorem**

2.36) 2.37)

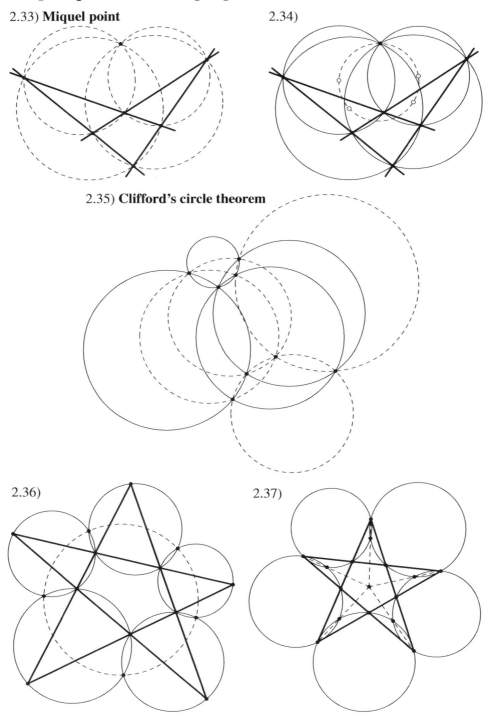

3 Triangle lines

3.1) **Euler line**

3.2) **Nagel line**

3.3)

3.4)

3.5)

3.6) **Soddy line**

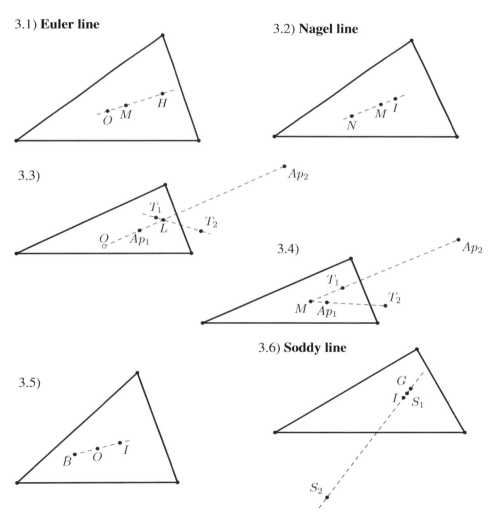

3.7) Aubert line

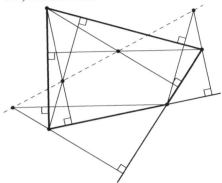

3.8) Gauss line

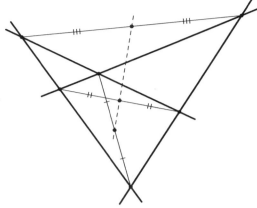

3.9)

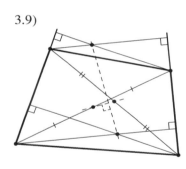

3.10) Plücker's theorem

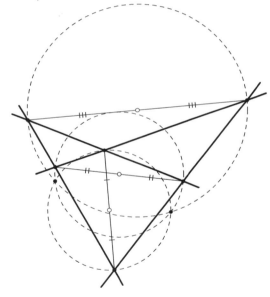

The Simson line and its propertiess

3.11) **Simson line**

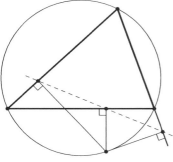

3.12) **General Simson line**

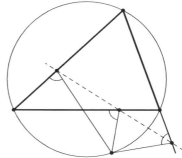

3.13)

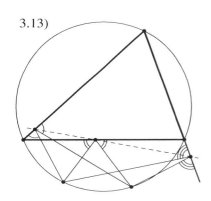

3.14)

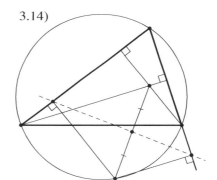

3.15)

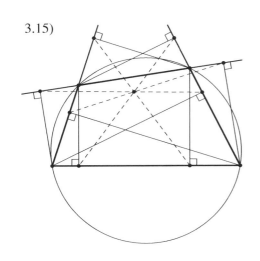

3.16)

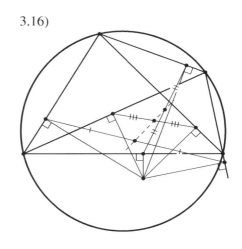

4 Elements of a triangle

4.1 Altitudes of a triangle

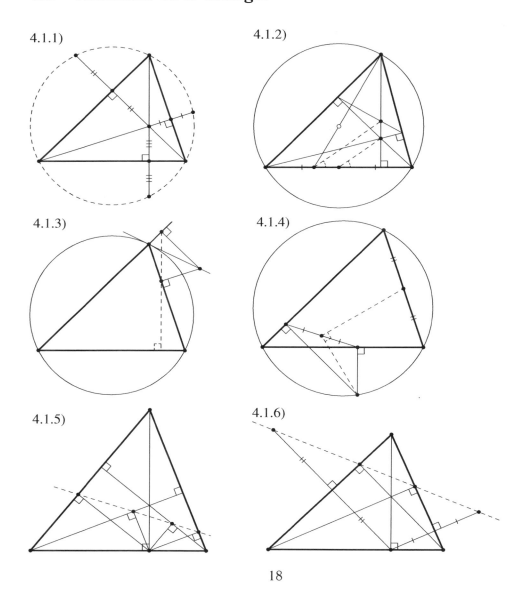

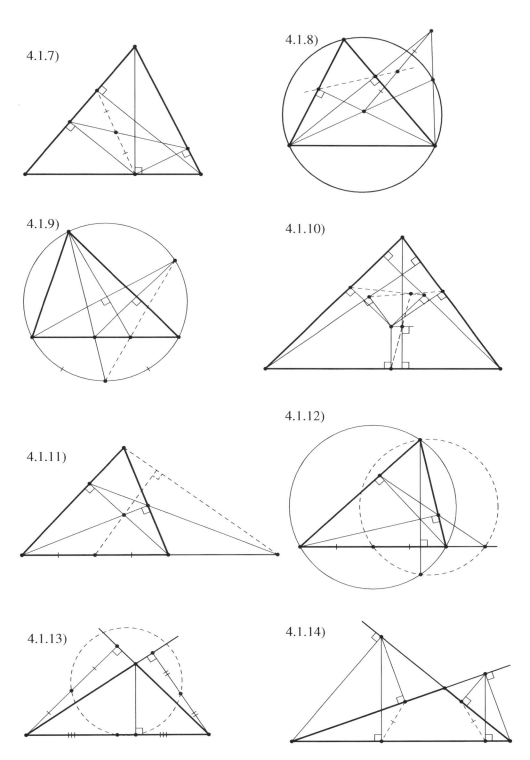

4.1.15)
4.1.16)
4.1.17)
4.1.18)
4.1.19)
4.1.20)
4.1.21)
4.1.22)
4.1.23)

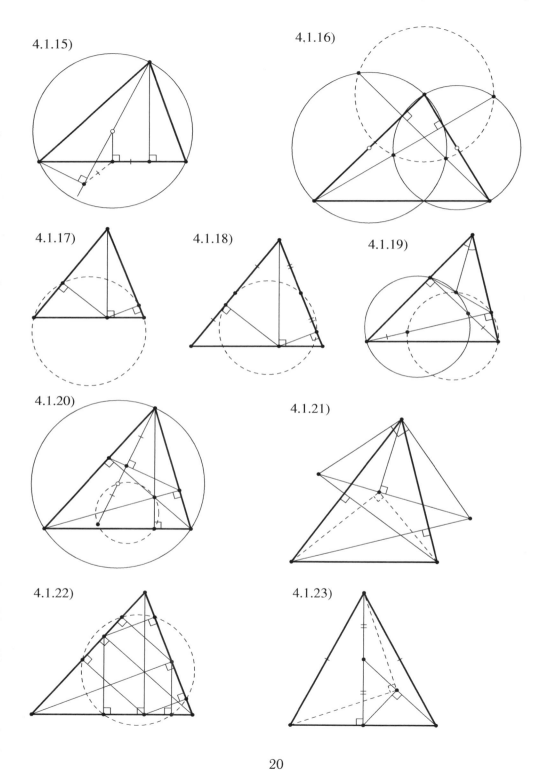

4.2 Orthocenter of a triangle

4.2.1)

4.2.2)

4.2.3)

4.2.4)

4.2.5)

4.2.6)

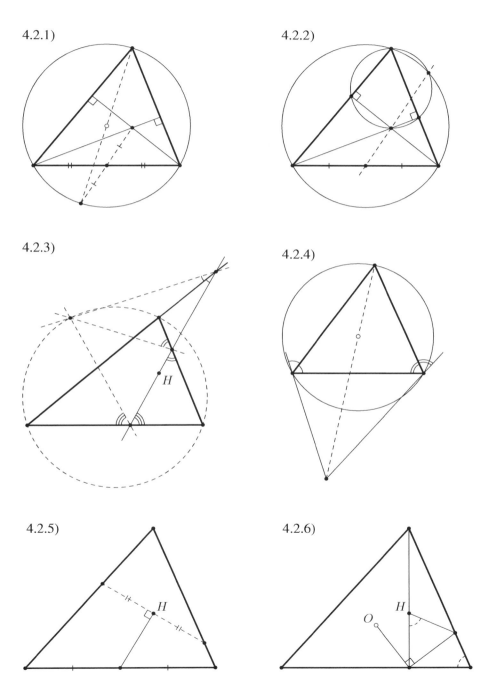

4.2.7) Droz-Farny theorem

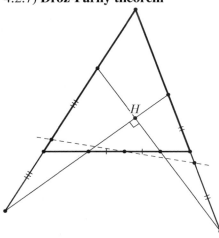

4.2.9)

4.2.8)

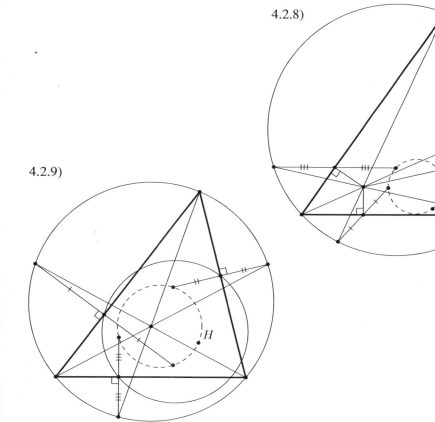

4.3 Angle bisectors of a triangle

4.3.1)

4.3.2)

4.3.3)

4.3.4)

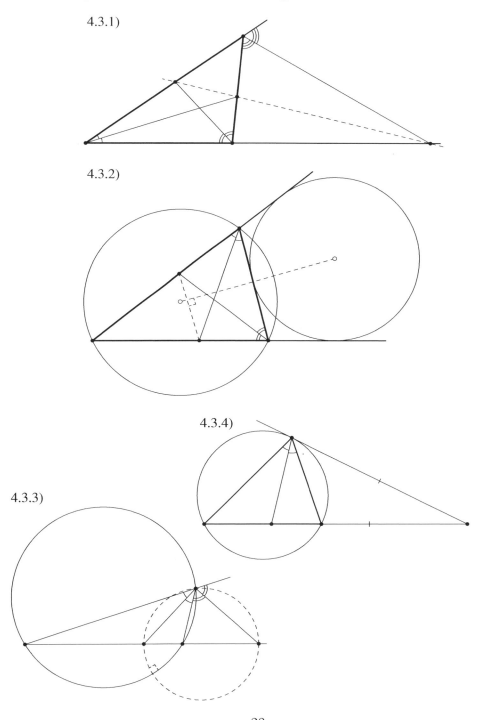

4.3.5)

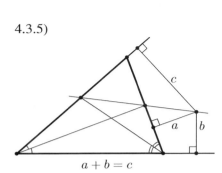

$a + b = c$

4.3.6)

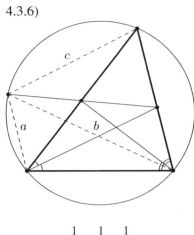

$$\frac{1}{a} = \frac{1}{b} + \frac{1}{c}$$

4.3.7)

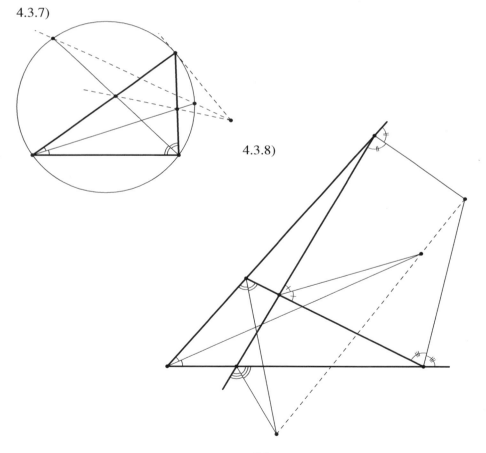

4.3.8)

4.3.9)
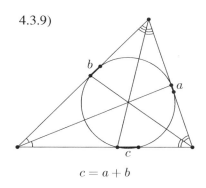
$c = a + b$

4.3.10)
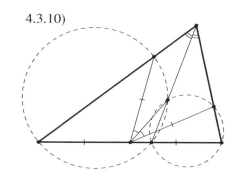

4.3.11)
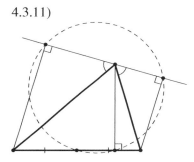

4.3.12)
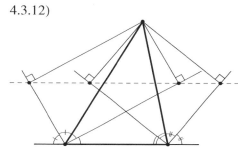

4.3.13)
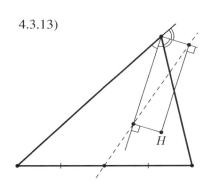

4.3.14)
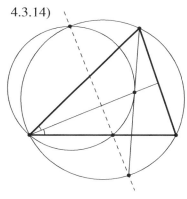

4.3.15)
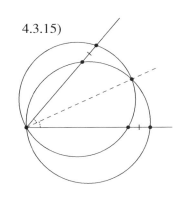

4.3.16)
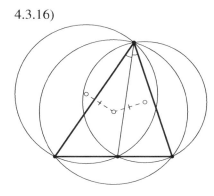

25

4.3.17)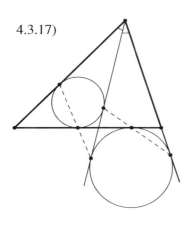

4.3.18)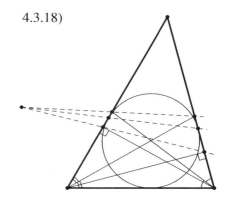

4.3.19)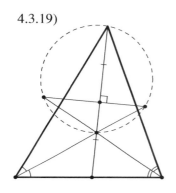

4.3.20)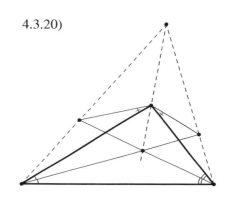

4.3.21)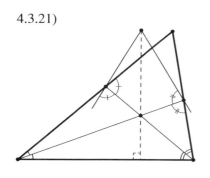

4.3.22)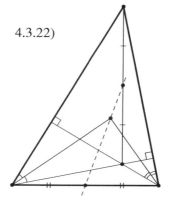

4.4 The symmedian and its properties

4.4.1)
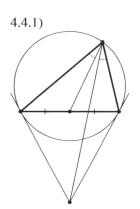

4.4.2)
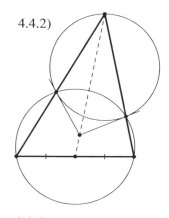

4.4.3)
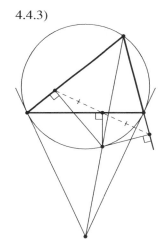

4.4.4)
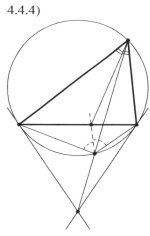

4.4.5)
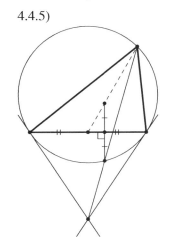

4.4.6)
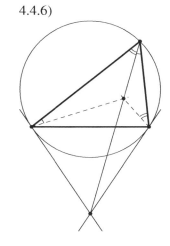

4.4.7)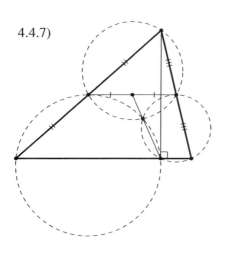

4.4.8)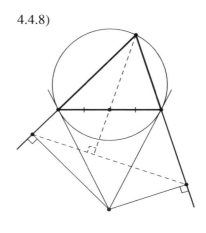

4.4.9)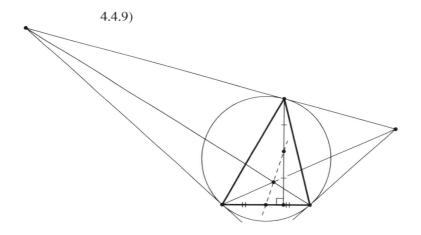

4.5 Inscribed circles

4.5.1)

4.5.2)

4.5.3)

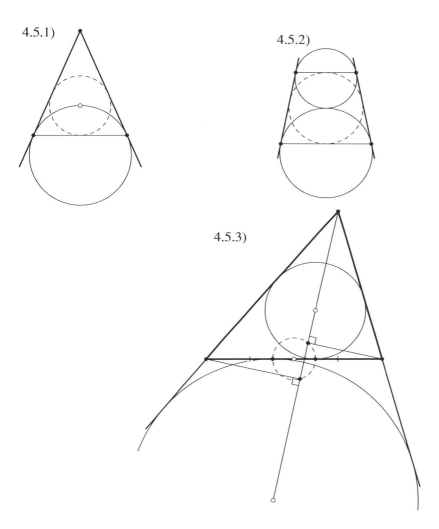

4.5.4)

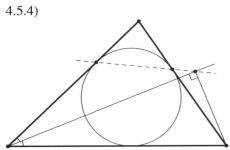

4.5.5)
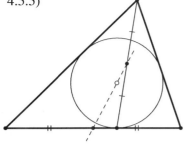

4.5.6)
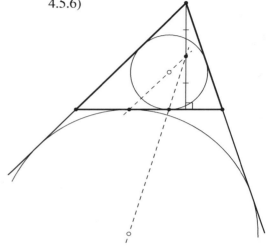

4.5.7)
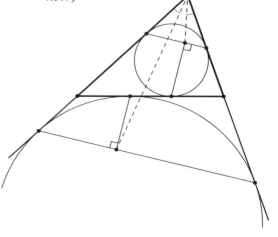

4.5.8)
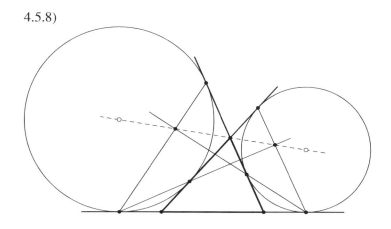

4.5.9)
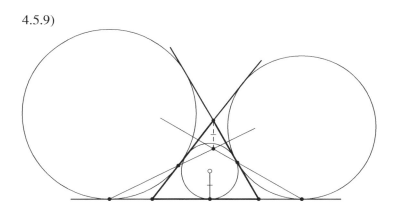

4.5.10)
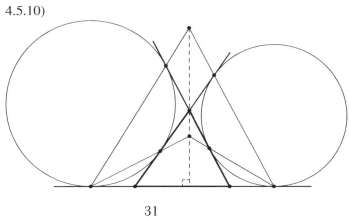

4.5.11)

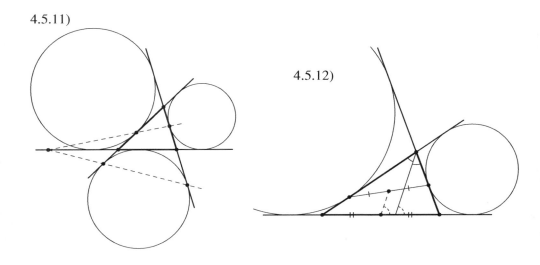

4.5.12)

4.5.13)

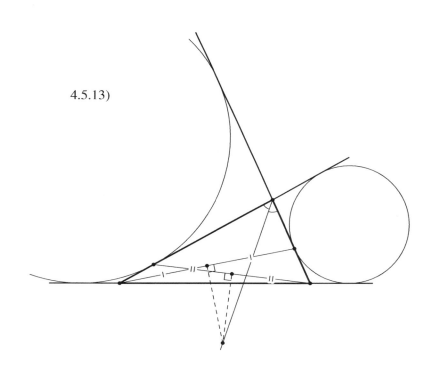

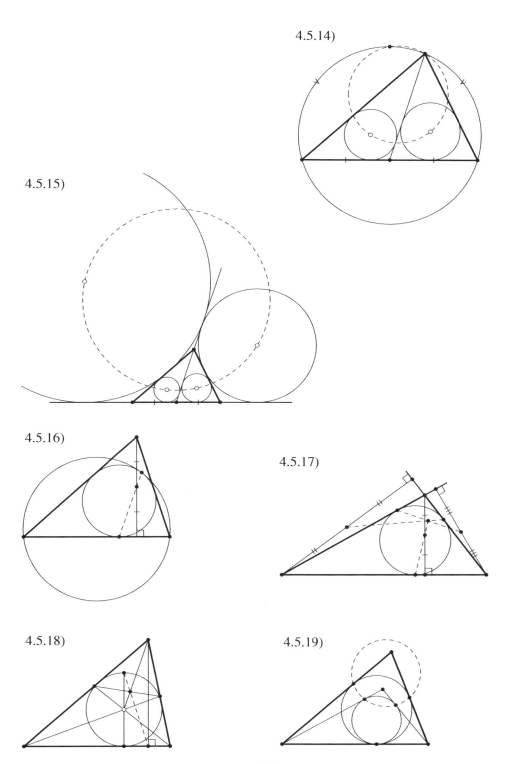

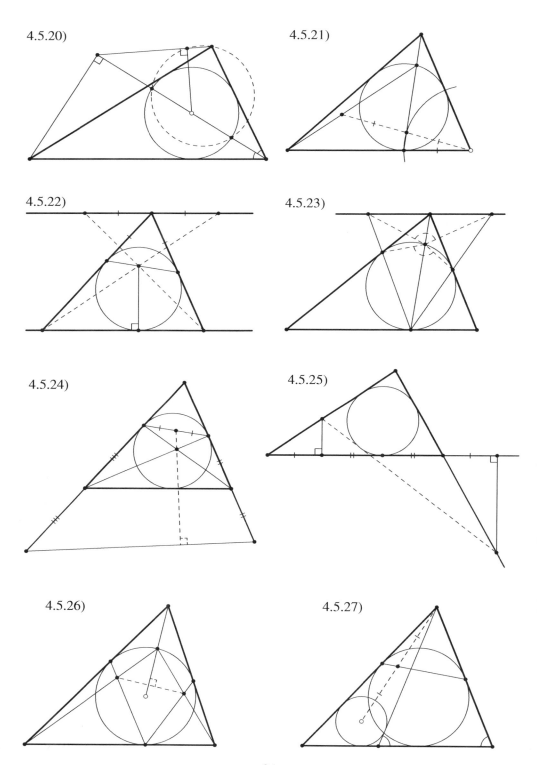

4.5.28) 4.5.29)

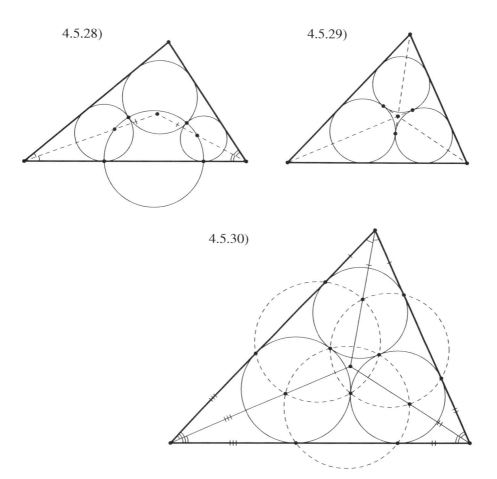

4.5.30)

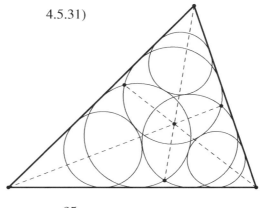

4.5.31)

4.5.32)

4.5.33)

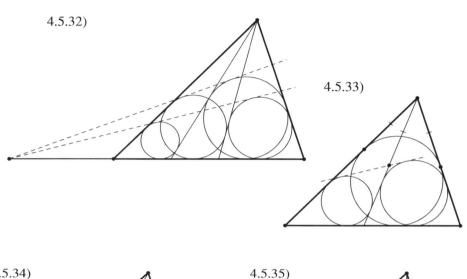

4.5.34)

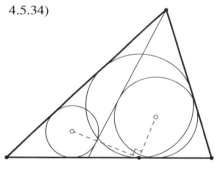

4.5.35)

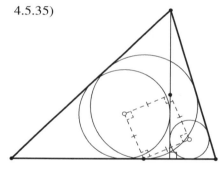

4.5.36)

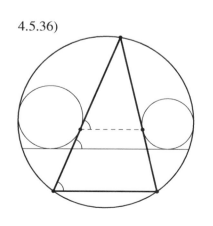

4.5.37)

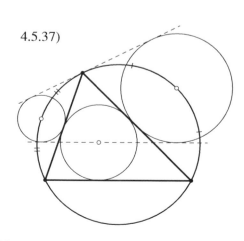

4.5.38)
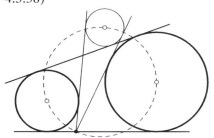

4.5.39)
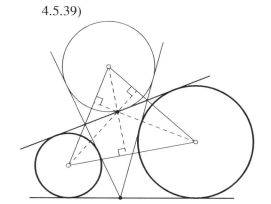

4.5.40)
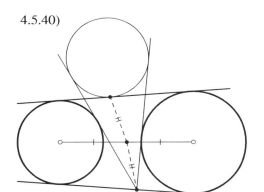

4.5.41)
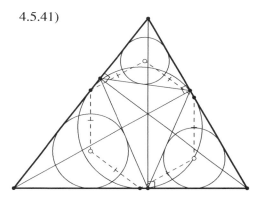

4.5.42)
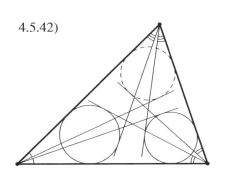

4.5.43)
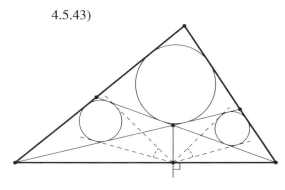

4.6 Inscribed and circumscribed circles of a triangle

4.6.1)

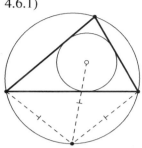

4.6.2)

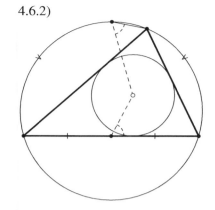

4.6.3) **Euler's formula**

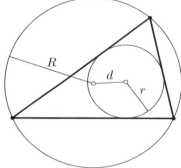

$$d^2 = R^2 - 2Rr$$

4.6.4)

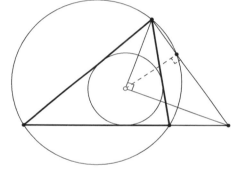

4.6.5)

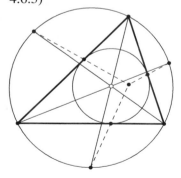

4.6.6)

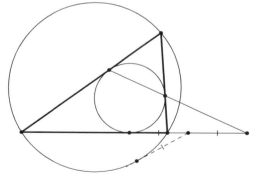

4.7 Circles tangent to the circumcircle of a triangle

Mixtilinear incircles

4.7.1) **Verrièr's lemma**

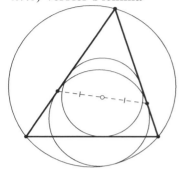

4.7.2)

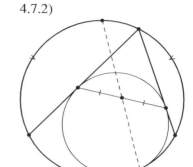

4.7.3)

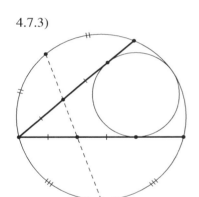

4.7.4)

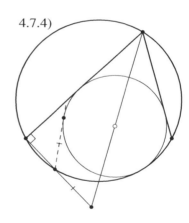

4.7.5)

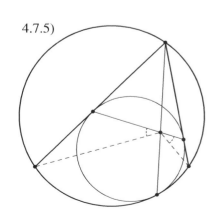

4.7.6)

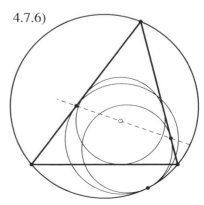

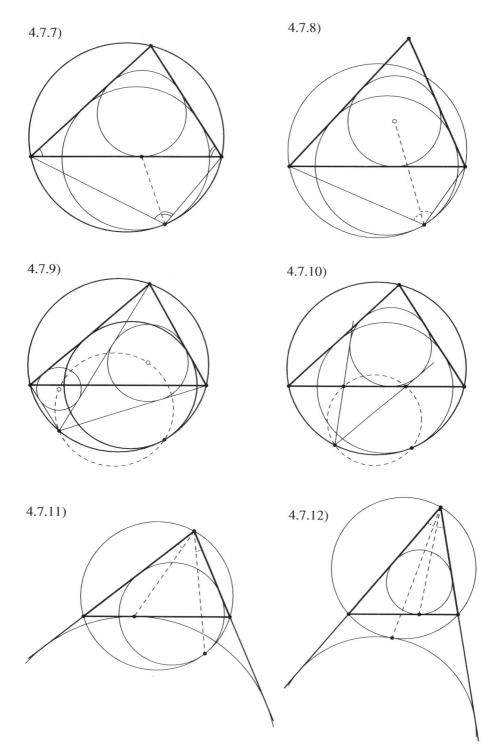

Segment theorem

4.7.13) **Sawayama's lemma**

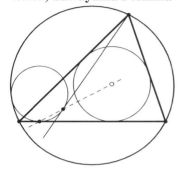

4.7.14)

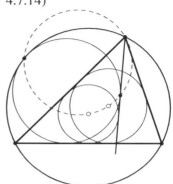

4.7.15) **Thébault's theorem**

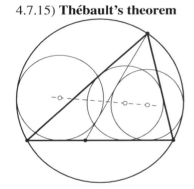

4.7.16)

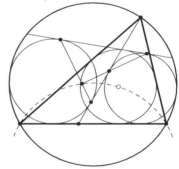

4.7.17)

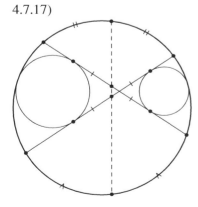

4.7.18)

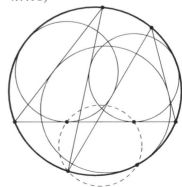

4.7.19)

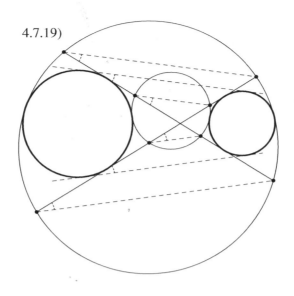

4.8 Circles related to a triangle

4.8.1) **Euler circle**

4.8.2) **Feuerbach's theorem**

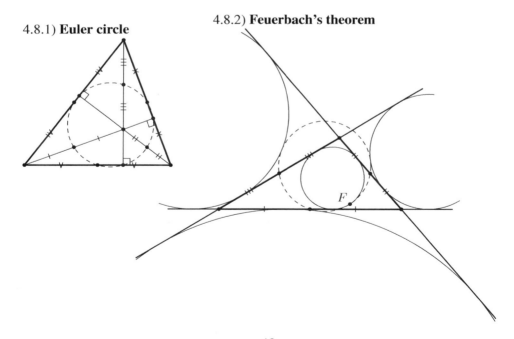

4.8.3) **Fontené's theorem**

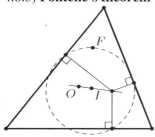

4.8.4) **Emelyanovs' theorem**

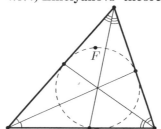

4.8.5)

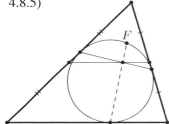

4.8.6)

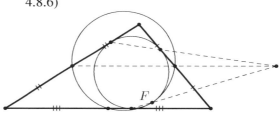

4.8.7)

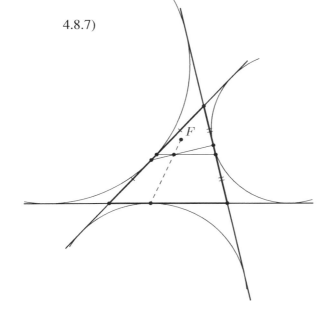

4.8.8)

4.8.9)

4.8.10) **Conway circle**

4.8.11) **van Lamoen circle**

4.8.12)

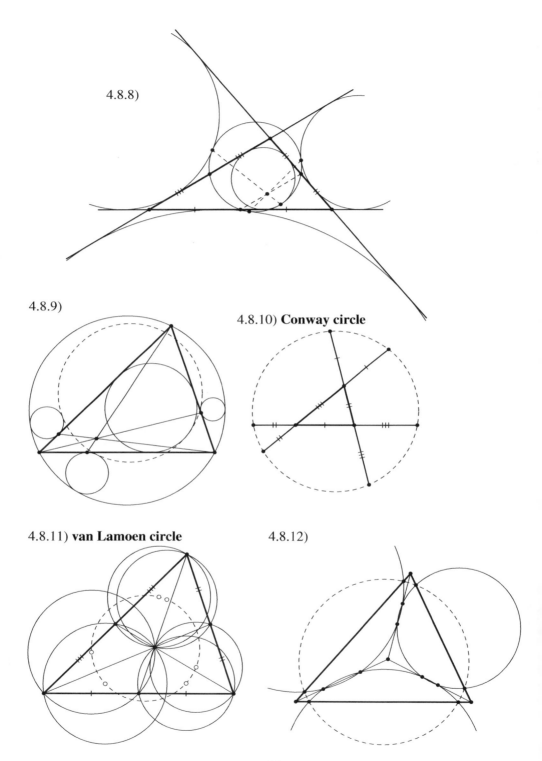

4.8.13)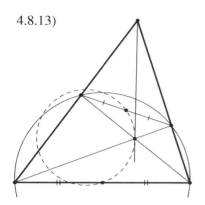

4.8.14)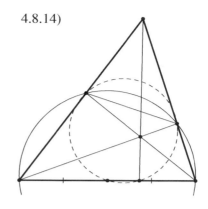

4.8.15)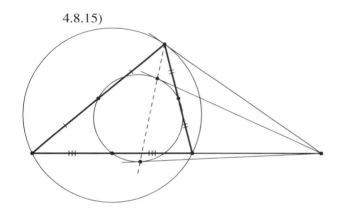

4.8.16)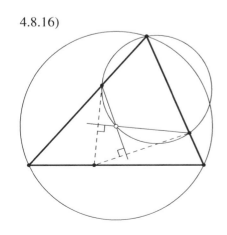

4.8.17)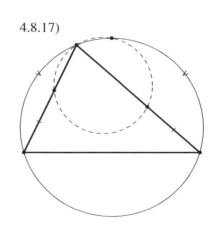

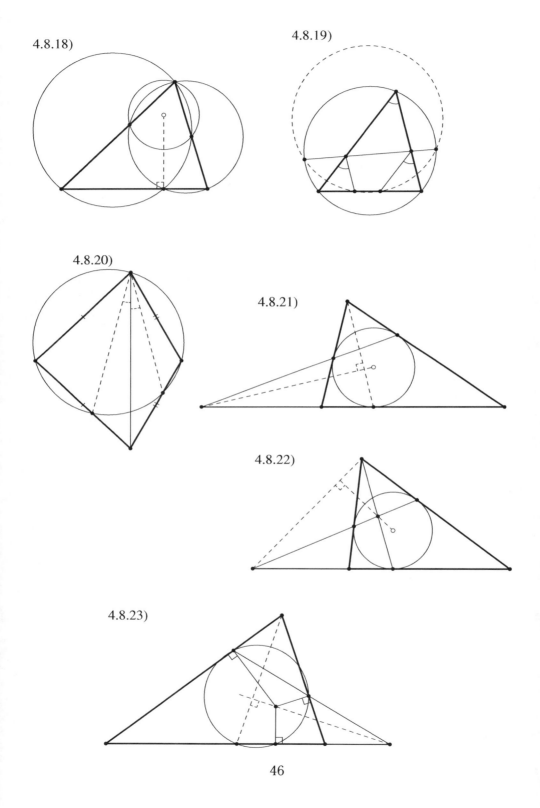

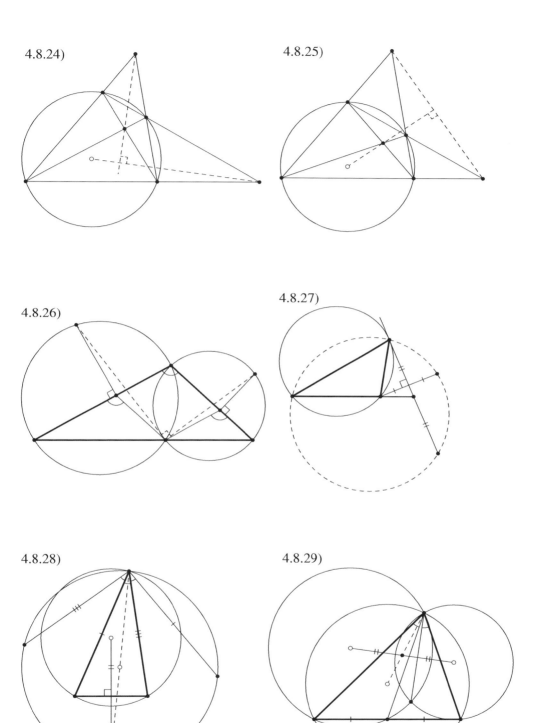

4.8.30)

4.8.31)

4.8.32)

4.8.33)

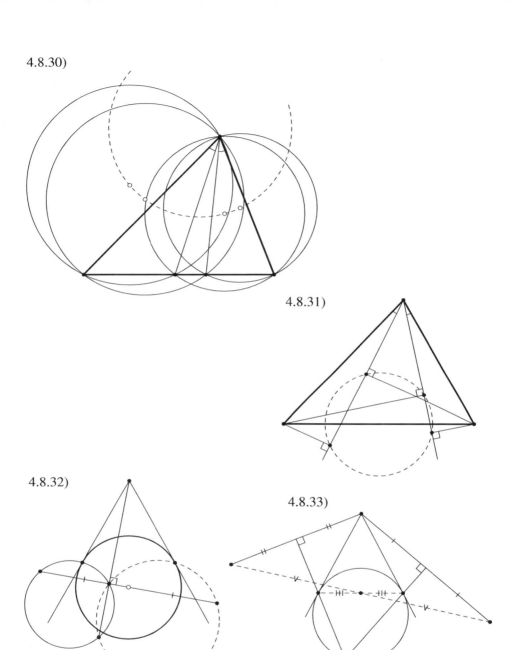

Common chord of two circles

4.8.34)

4.8.35)

4.8.36)

4.8.37)

4.8.38)

4.8.39)

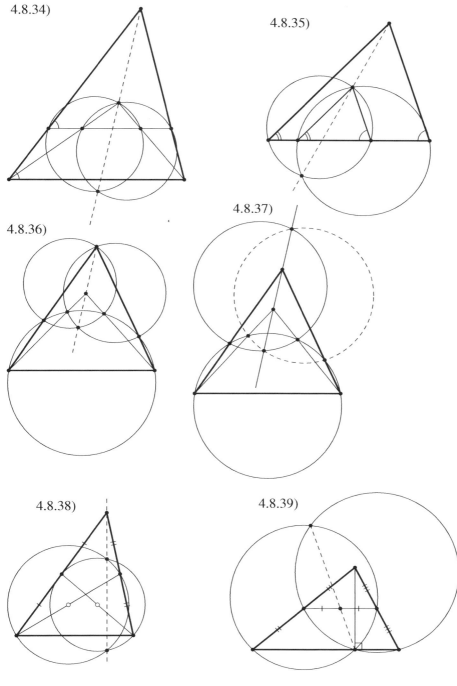

4.9 Concurrent lines of a triangle

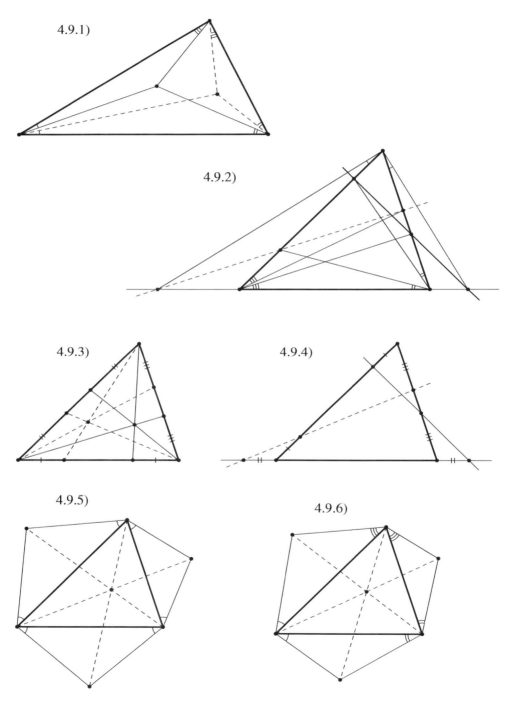

4.9.1)

4.9.2)

4.9.3)

4.9.4)

4.9.5)

4.9.6)

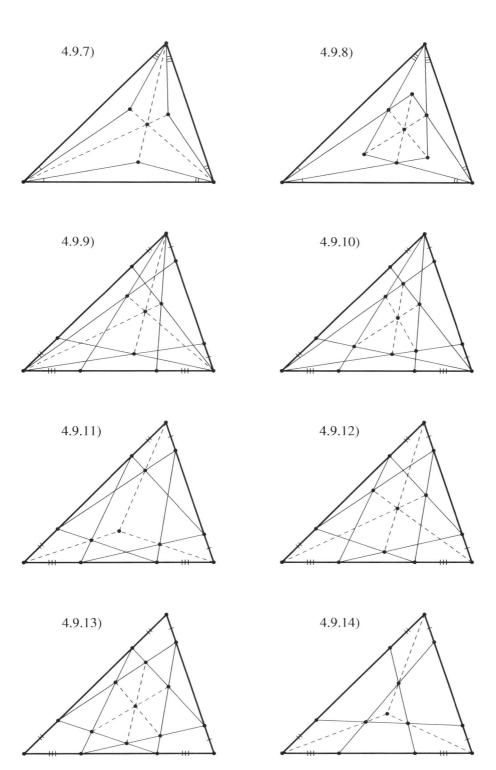

4.9.15) **Ceva's theorem**

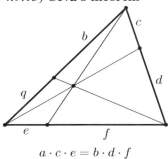

$$a \cdot c \cdot e = b \cdot d \cdot f$$

4.9.16) **Carnot's theorem**

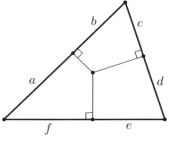

$$a^2 + c^2 + e^2 = b^2 + d^2 + f^2$$

4.9.17)

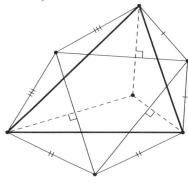

4.9.18) **Steiner's theorem**

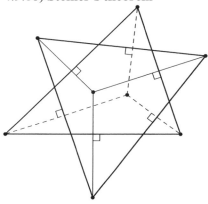

4.9.19)

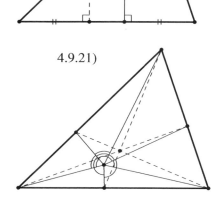

4.9.20)

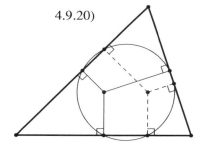

4.9.21)

4.9.22)

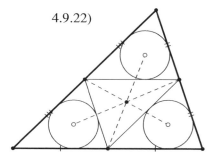

4.9.23)

4.9.24)

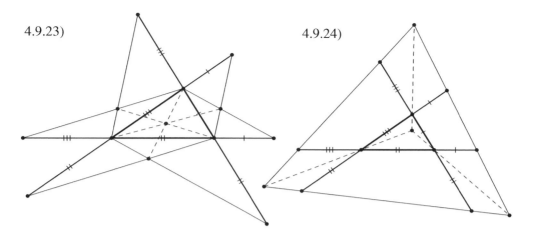

4.9.25)

4.9.26)

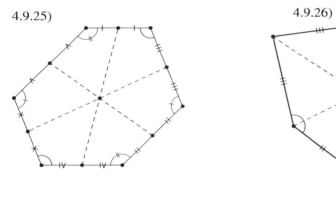

4.9.27)

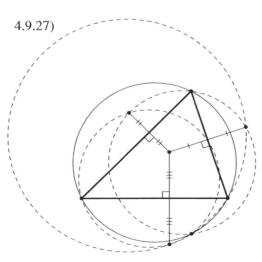

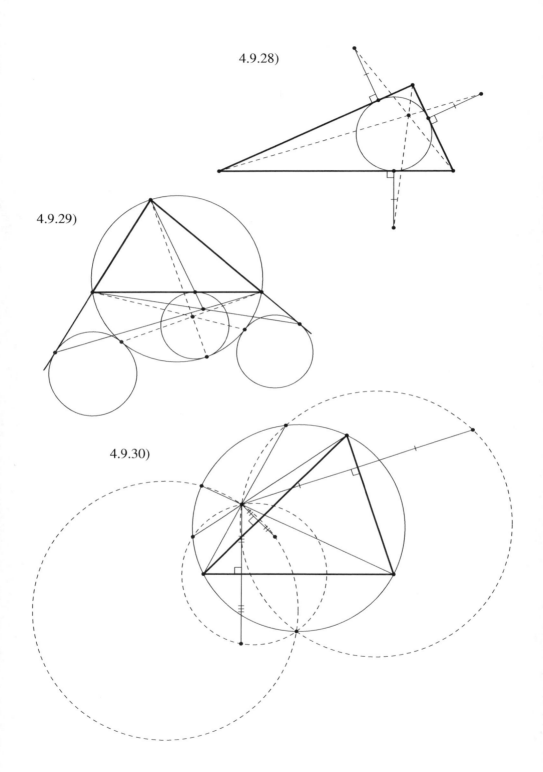

4.10 Right triangles

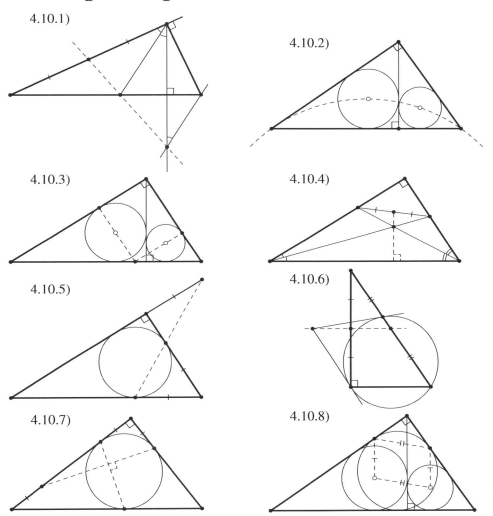

4.10.1)
4.10.2)
4.10.3)
4.10.4)
4.10.5)
4.10.6)
4.10.7)
4.10.8)

4.11 Theorems about certain angles

4.11.1)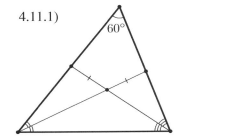

4.11.2)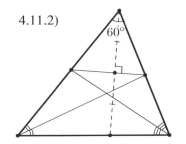

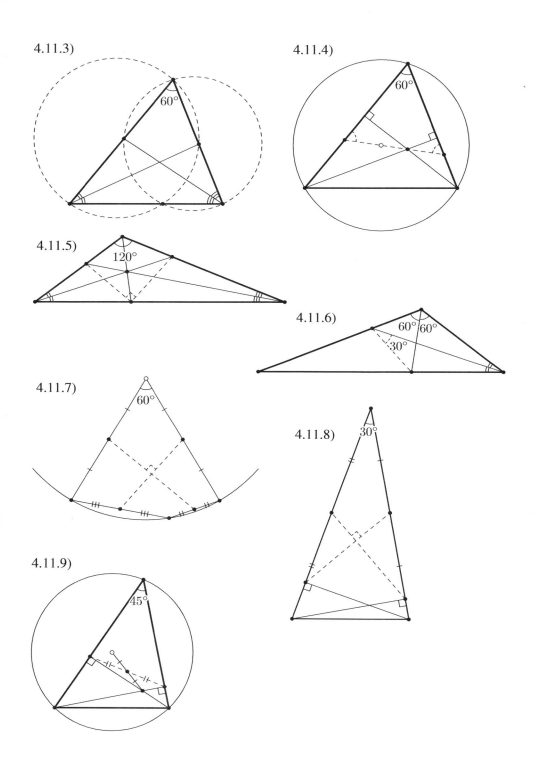

4.12 Other problems and theorems

4.12.1) **Blanchet's theorem**

4.12.2)

4.12.3)

4.12.4)

4.12.5)

4.12.6)

4.12.7)

4.12.8)

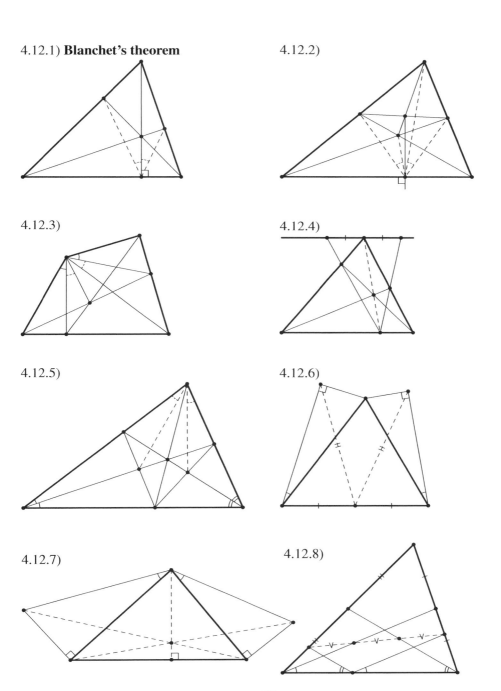

4.12.9)

4.12.10)

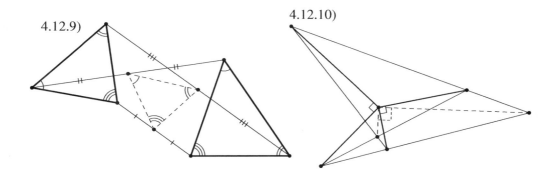

4.12.11) **Morley's theorem**

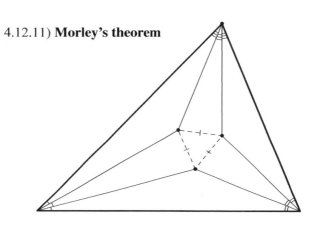

4.12.12)

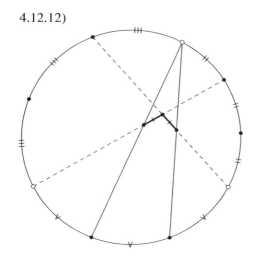

5 Quadrilaterals

5.1 Parallelograms

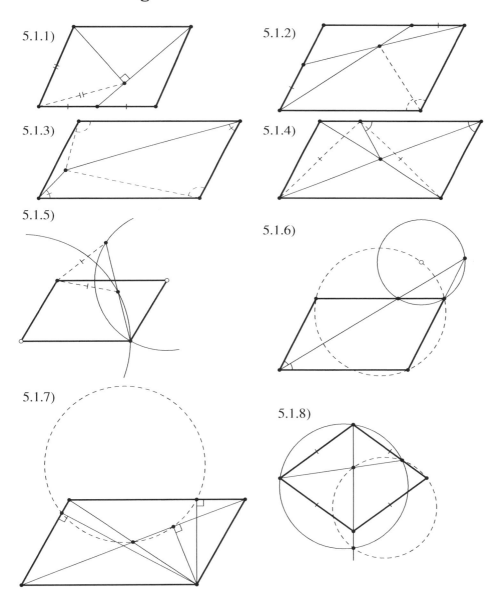

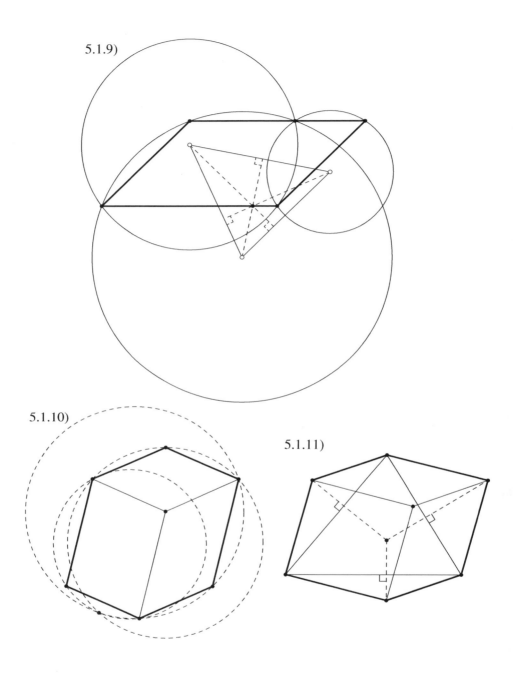

5.2 Trapezoids

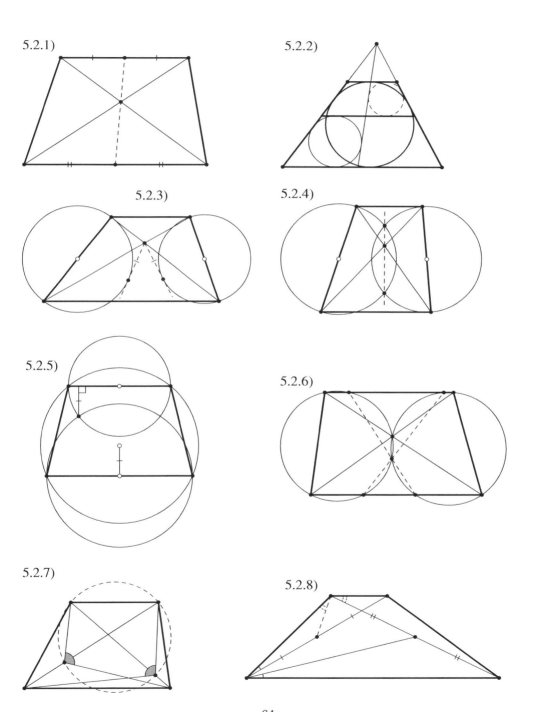

5.2.9)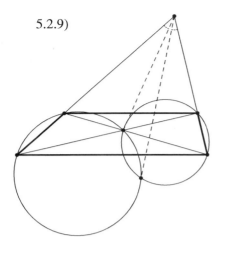

5.2.10)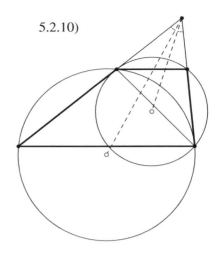

5.3 Squares

5.3.1)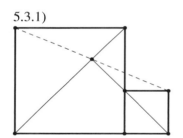

5.3.2)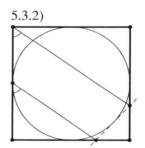

5.3.3)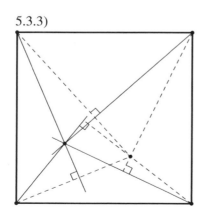

5.4 Circumscribed quadrilaterals

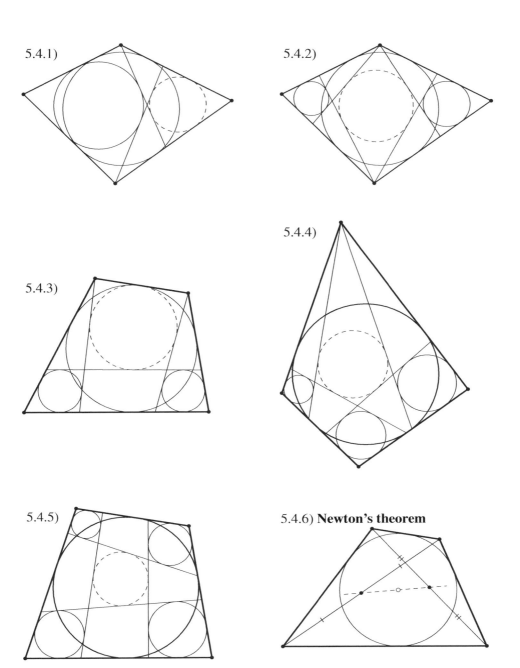

5.4.1)

5.4.2)

5.4.3)

5.4.4)

5.4.5)

5.4.6) **Newton's theorem**

5.4.7)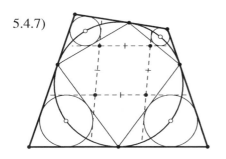

5.4.8)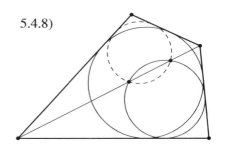

5.4.9)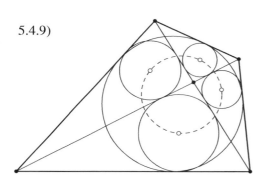

5.4.10)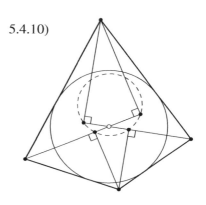

5.4.11)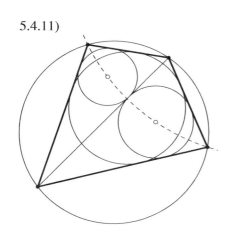

5.4.12)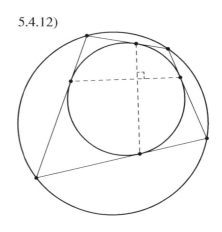

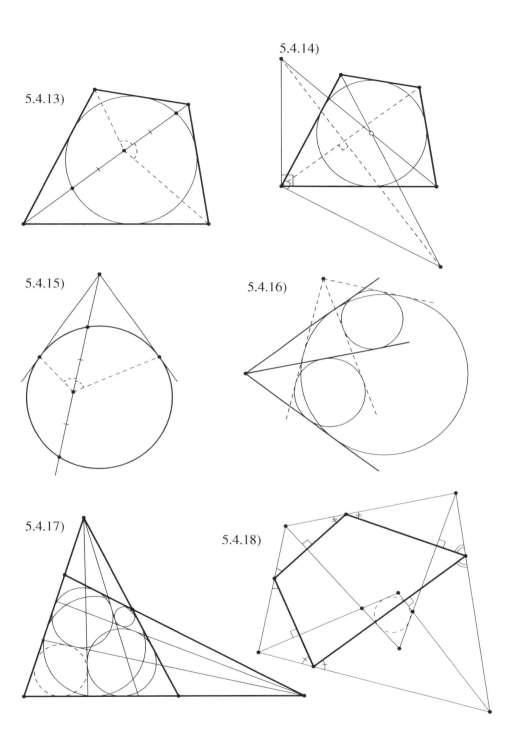

5.5 Inscribed quadrilaterals

5.5.1)

5.5.2)

5.5.3)

5.5.4)

5.5.5)

5.5.6) **Ptolemy's theorem**

$a \cdot c + b \cdot d = e \cdot f$

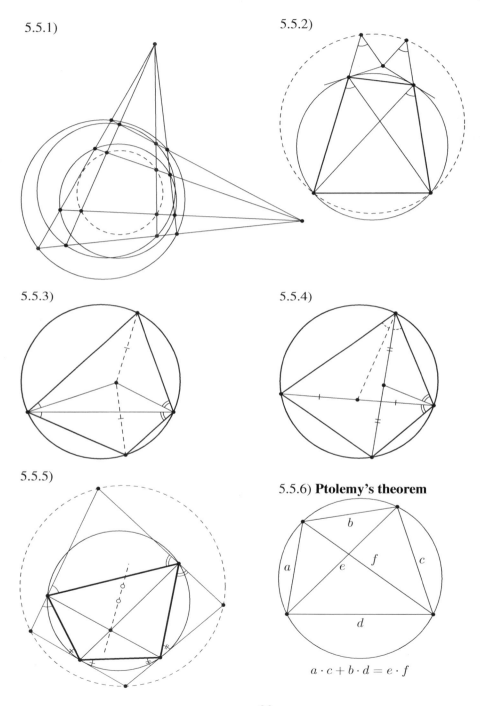

5.5.7)
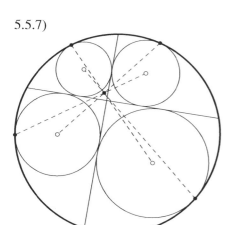

5.5.8)
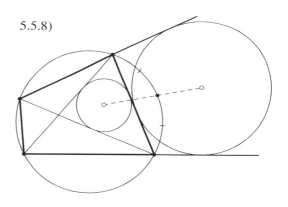

5.5.9)
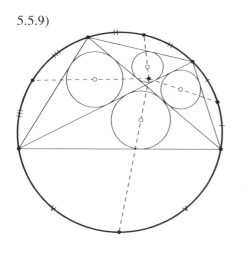

5.5.10)
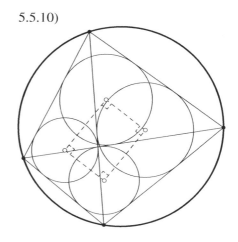

5.6 Four points on a circle

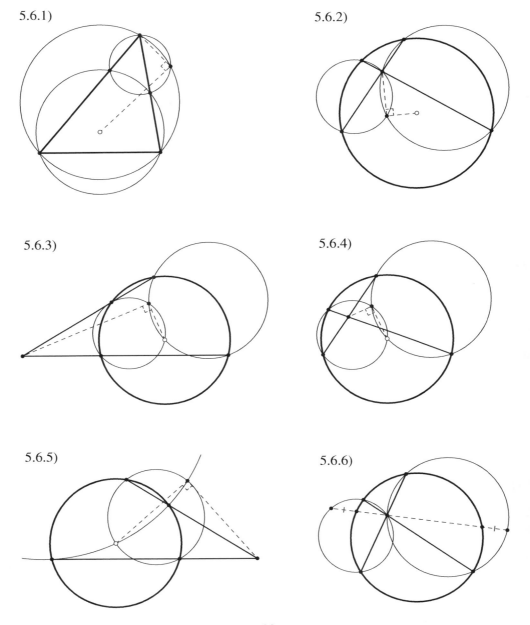

5.6.1)

5.6.2)

5.6.3)

5.6.4)

5.6.5)

5.6.6)

5.6.7)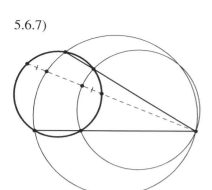

5.6.8)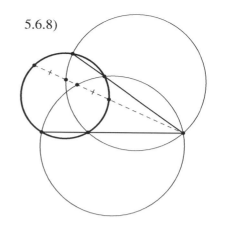

5.6.9)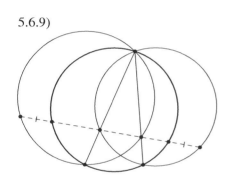

5.6.10)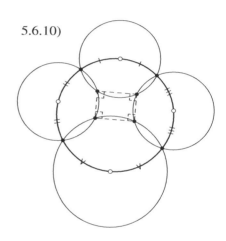

5.6.11)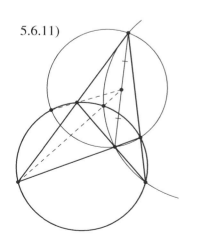

5.6.12)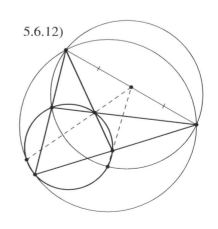

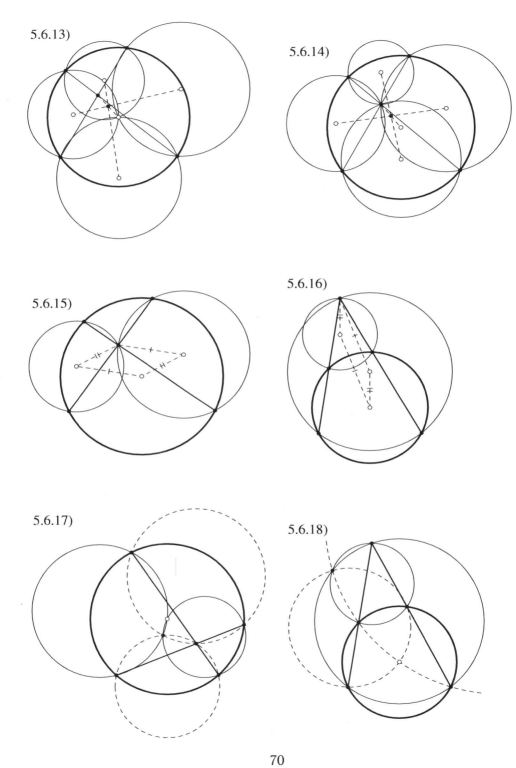

5.7 Altitudes in quadrilaterals

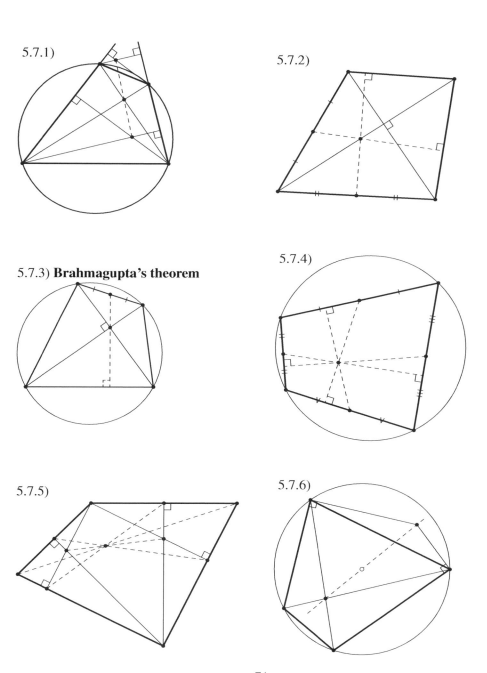

5.7.1)

5.7.2)

5.7.3) **Brahmagupta's theorem**

5.7.4)

5.7.5)

5.7.6)

5.7.7)

5.7.8)

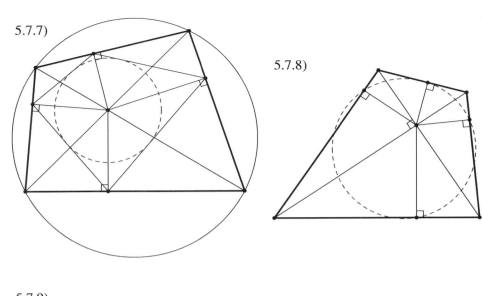

5.7.9)

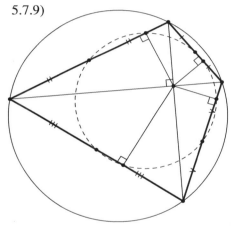

5.7.10)

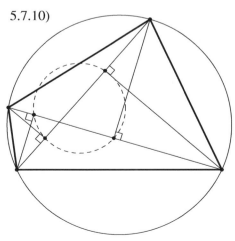

6 Circles

6.1 Tangent circles

6.1.1)

6.1.2)

6.1.3)

6.1.4)

6.1.5)

6.1.6)

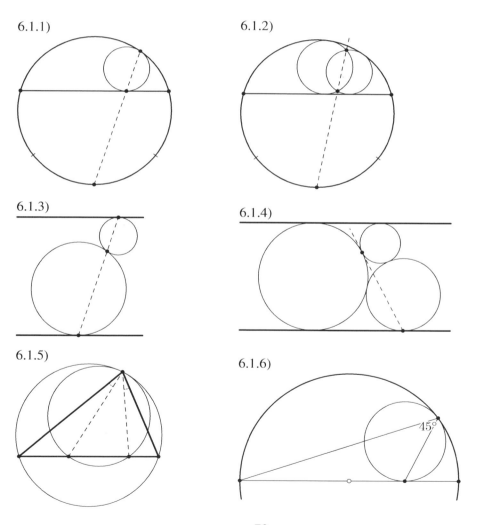

6.1.7)
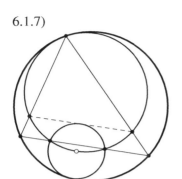

6.1.8)
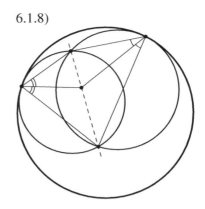

6.1.9)
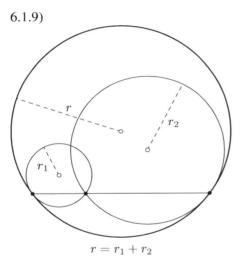

$r = r_1 + r_2$

6.1.10) **Casey's theorem**
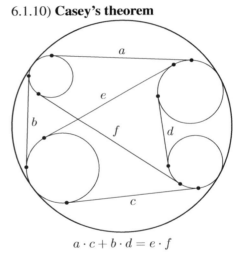

$a \cdot c + b \cdot d = e \cdot f$

6.2 Monge's theorem and related constructions

6.2.1) Eyeball theorem

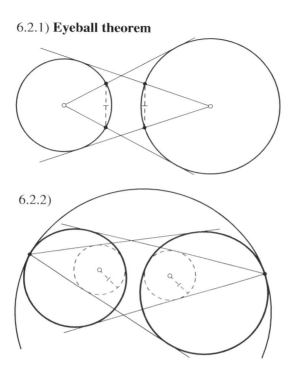

6.2.2)

6.2.3) Monge's theorem

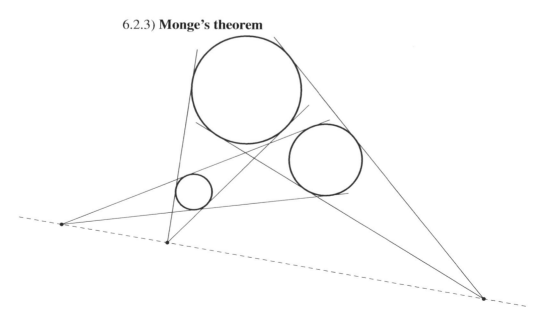

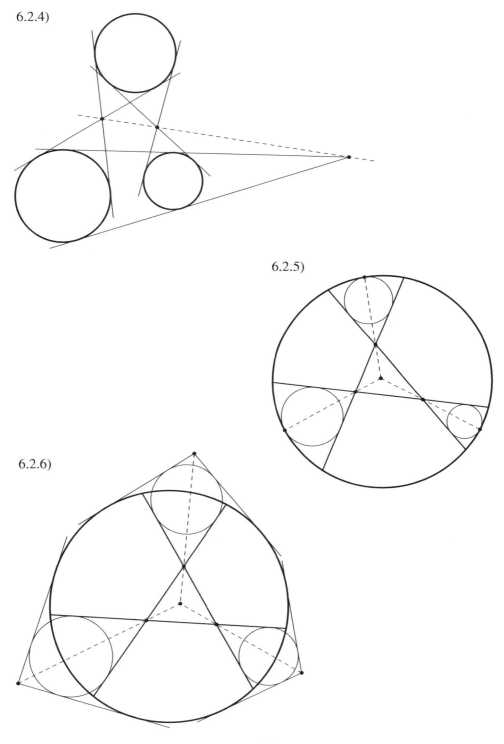

6.2.7)

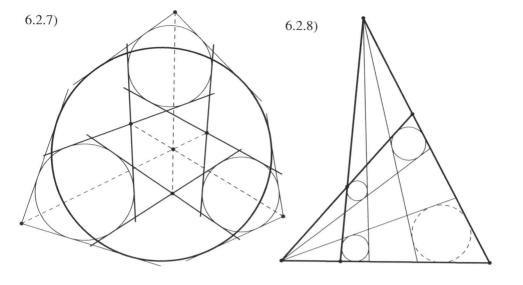

6.2.8)

6.2.9)

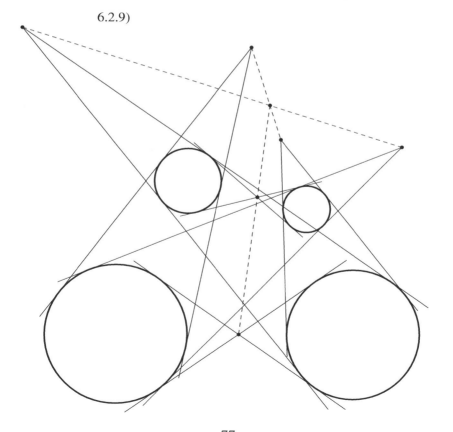

6.2.10)

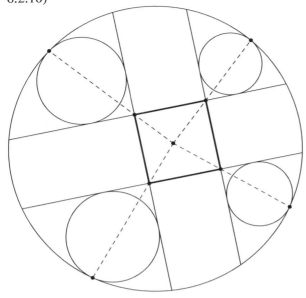

6.2.11)

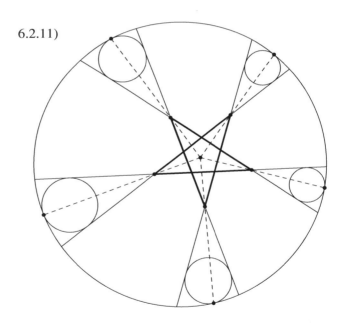

6.3 Common tangents of three circles

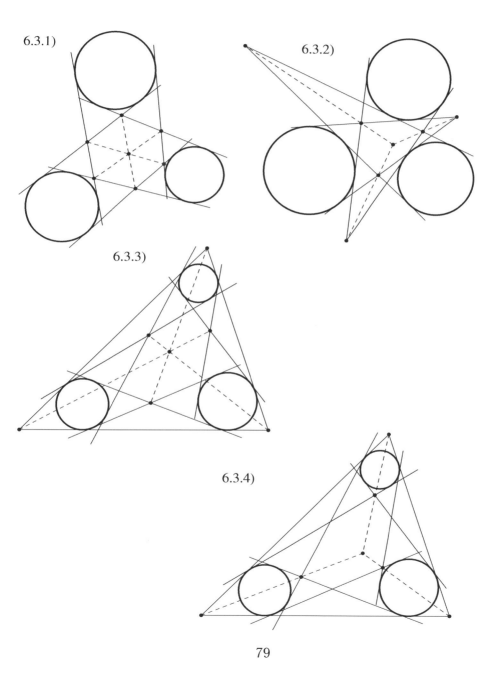

6.3.1)

6.3.2)

6.3.3)

6.3.4)

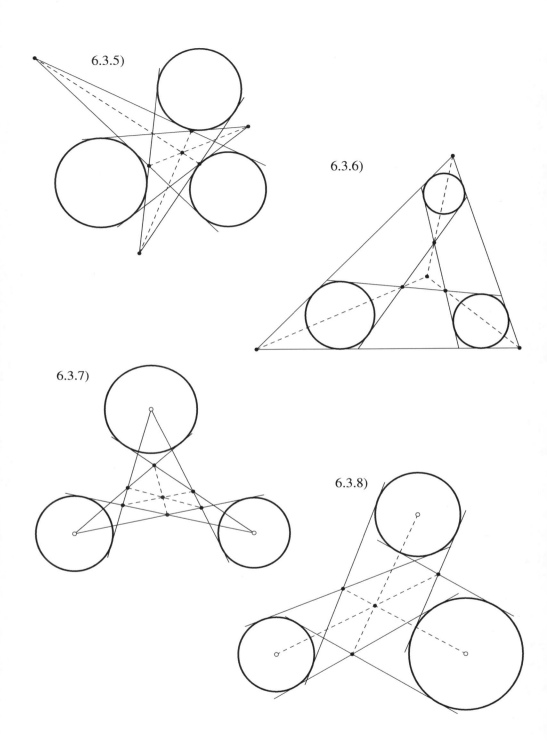

6.4 Butterfly theorem

6.4.1)

6.4.2)

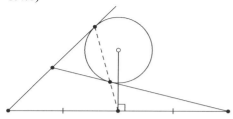
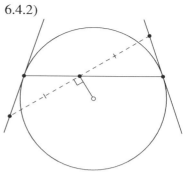

6.4.3) **Butterfly theorem**

6.4.4) **Dual butterfly theorem**

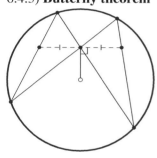
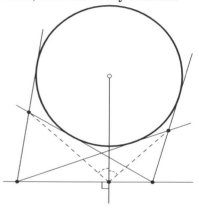

6.4.5)

6.4.6)

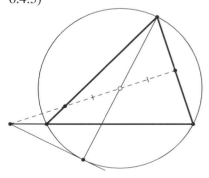
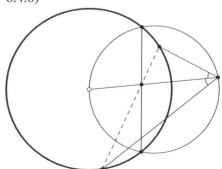

6.5 Power of a point and related questions

6.5.1) **Radical axis theorem**

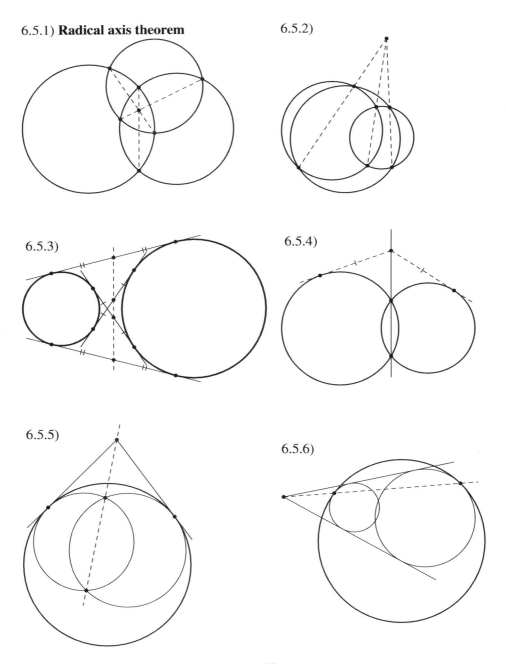

6.5.2)

6.5.3)

6.5.4)

6.5.5)

6.5.6)

6.5.7)

6.5.8)

6.5.9)

6.5.10)

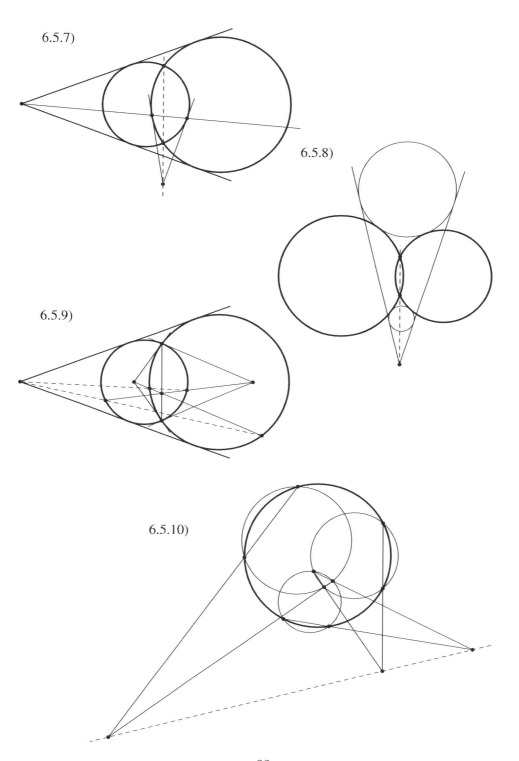

6.6 Equal circles

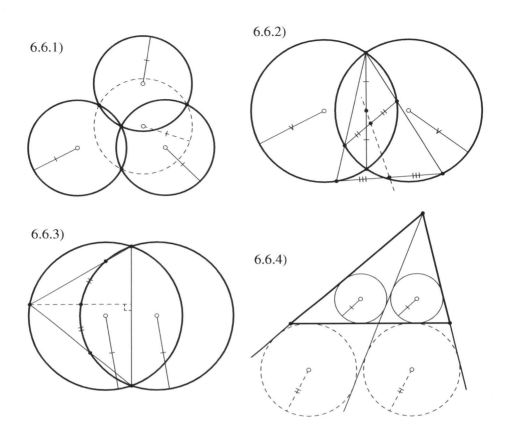

6.7 Diameter of a circle

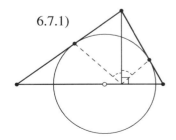
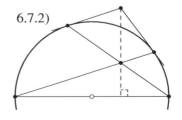

6.7.3)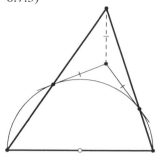

6.7.4)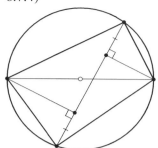

6.7.5)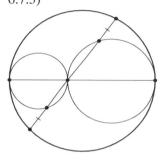

6.7.6)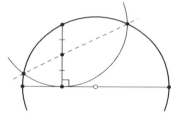

6.7.7)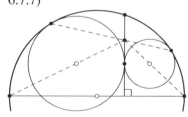

6.7.8)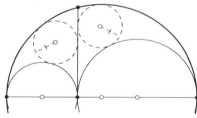

6.7.9)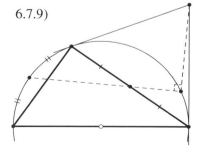

6.8 Constructions from circles

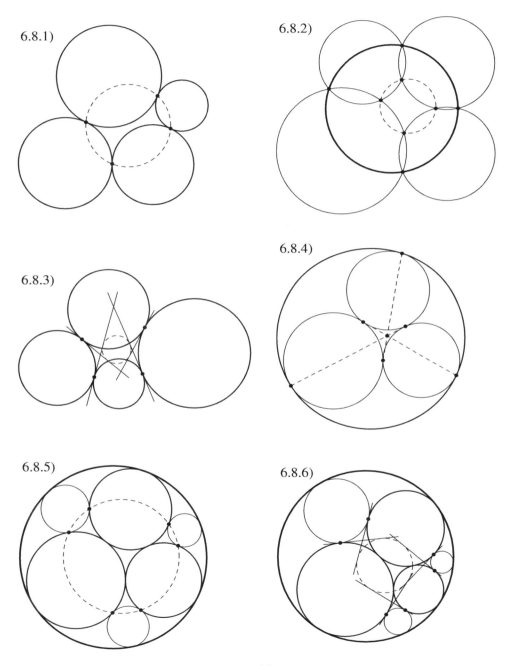

6.8.1)

6.8.2)

6.8.3)

6.8.4)

6.8.5)

6.8.6)

6.8.7)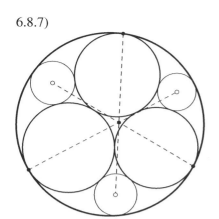

6.8.8)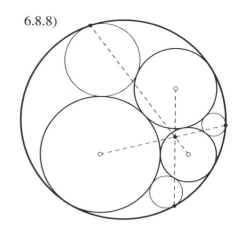

6.8.9)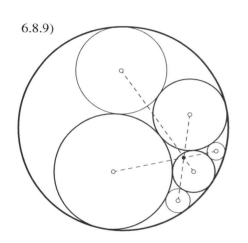

6.8.10) **Seven circles theorem**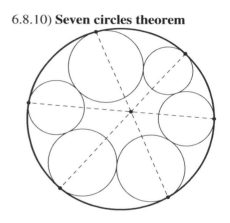

6.8.11)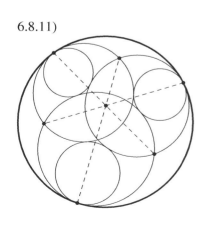

6.8.12)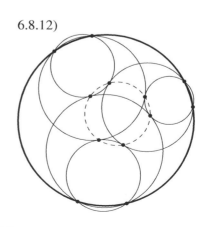

6.8.13)

6.8.14)
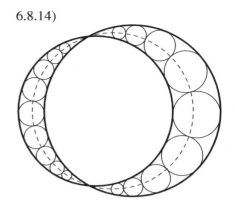

6.8.15) **Hart's theorem**
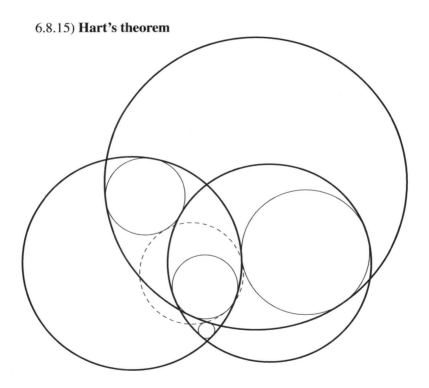

6.9 Circles tangent to lines

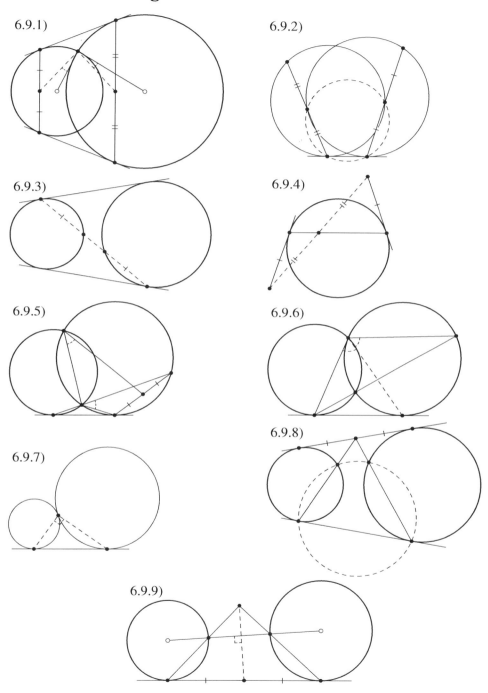

6.9.1)
6.9.2)
6.9.3)
6.9.4)
6.9.5)
6.9.6)
6.9.7)
6.9.8)
6.9.9)

6.10 Miscellaneous problems

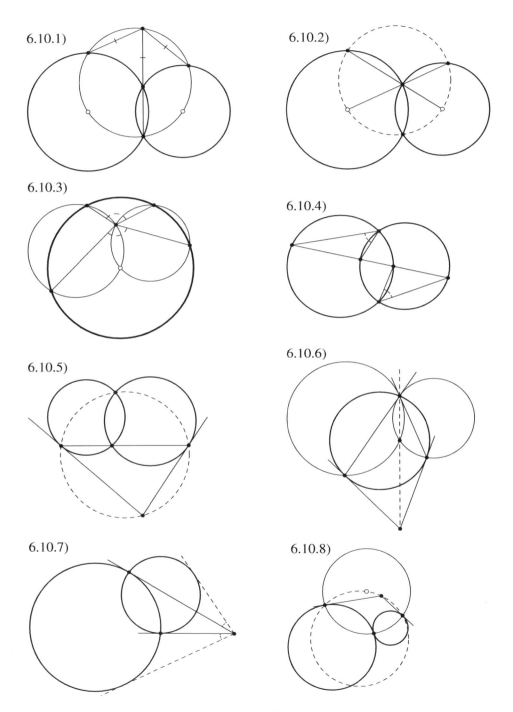

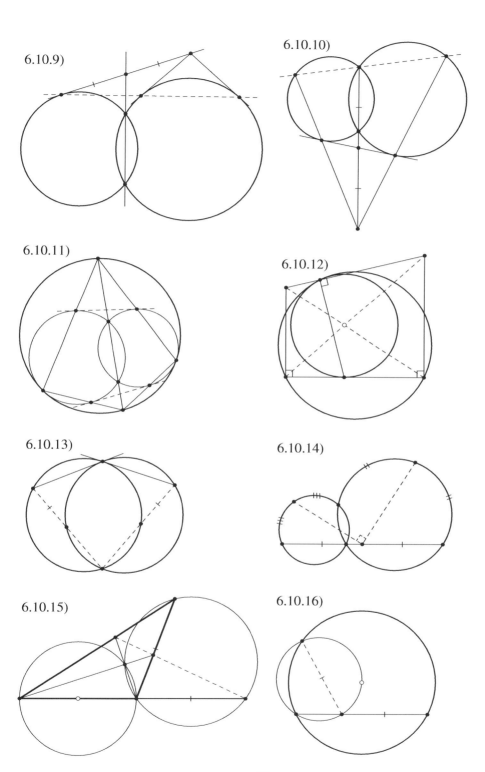

6.10.17)
6.10.18)
6.10.19)
6.10.20)
6.10.21)
6.10.22)
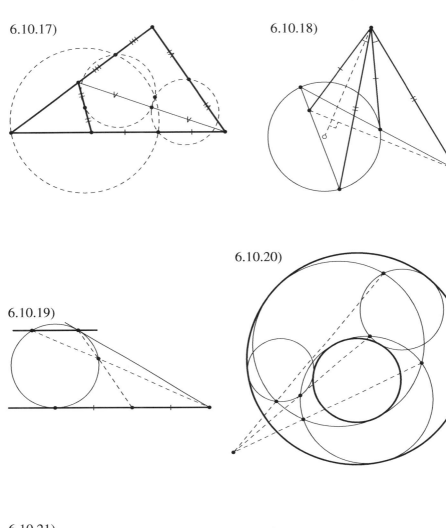
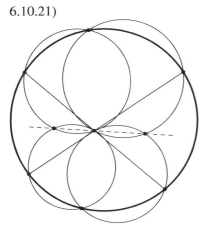
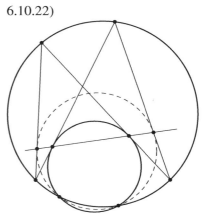

6.10.23)

6.10.24)

$a \cdot c \cdot e = b \cdot d \cdot f$

6.10.25)

$a \cdot c \cdot e = b \cdot d \cdot f$

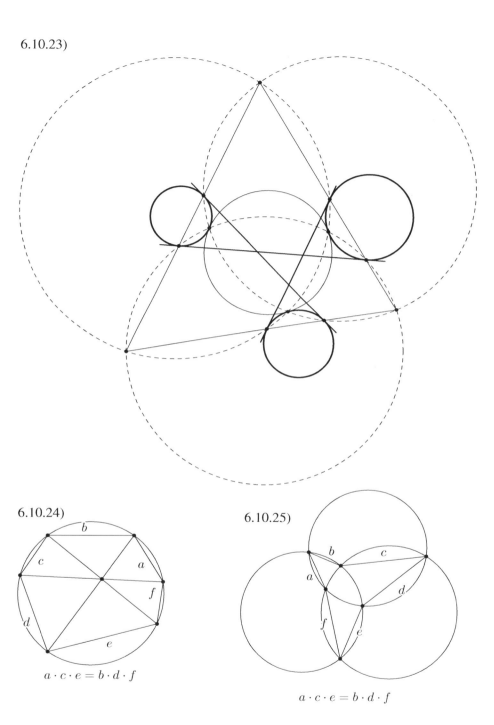

7 Projective theorems

7.1) **Desargues' theorem**

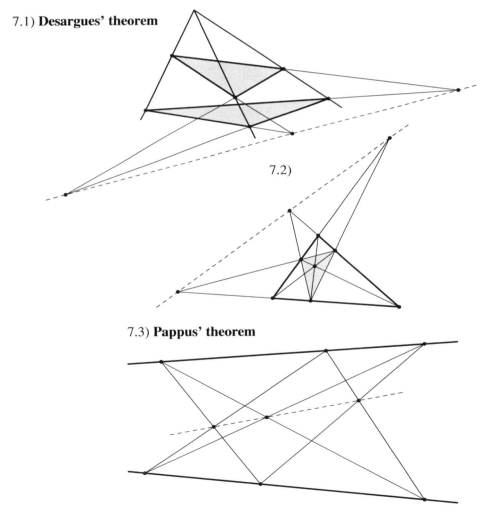

7.2)

7.3) **Pappus' theorem**

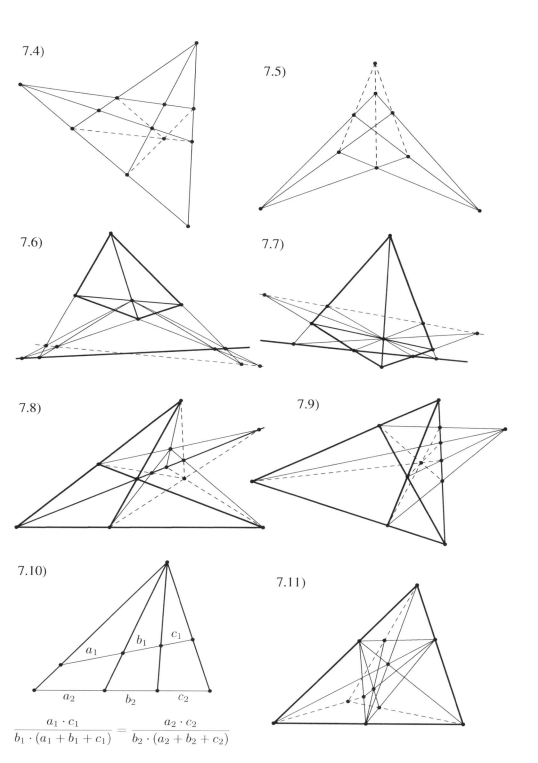

$$\frac{a_1 \cdot c_1}{b_1 \cdot (a_1 + b_1 + c_1)} = \frac{a_2 \cdot c_2}{b_2 \cdot (a_2 + b_2 + c_2)}$$

8 Regular polygons

8.1)

8.2)

8.3)

8.4)

8.5)

8.6)

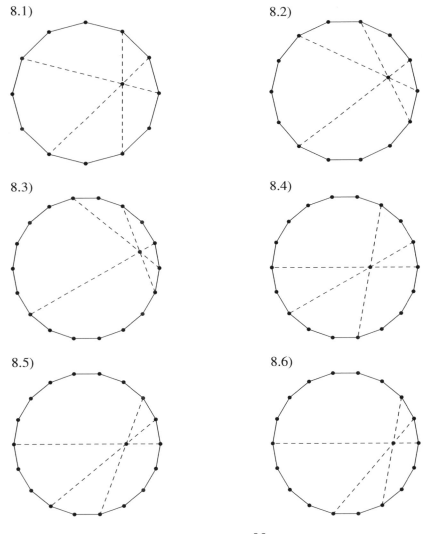

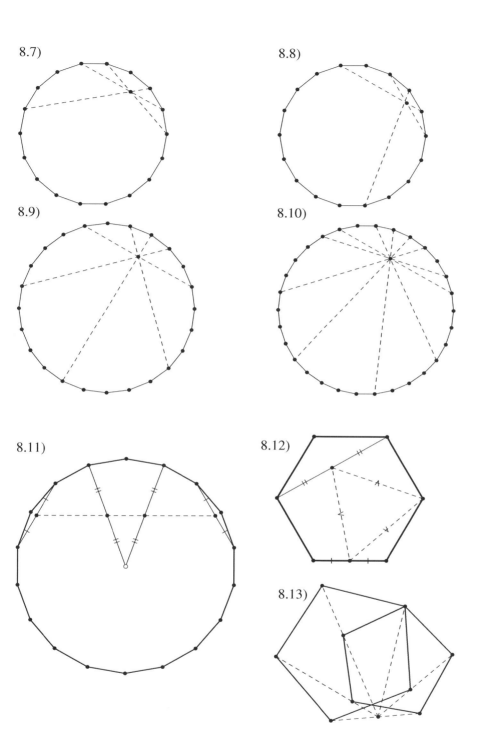

8.1 Remarkable properties of the equilateral triangle

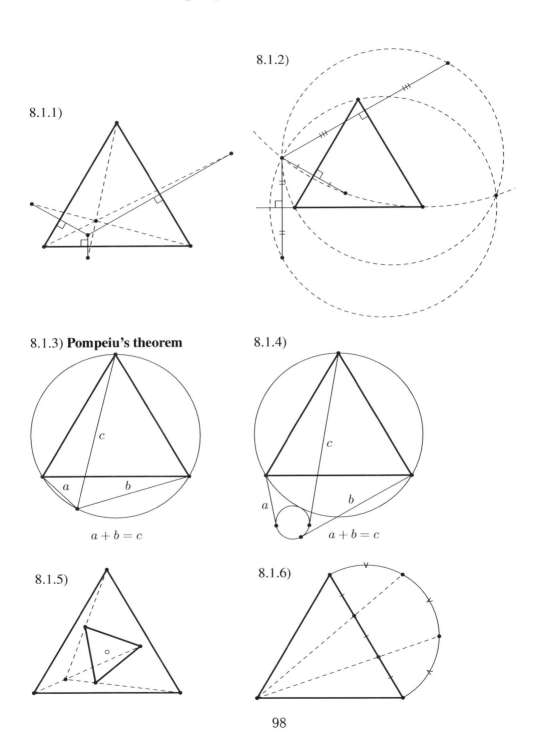

8.1.1)

8.1.2)

8.1.3) **Pompeiu's theorem**

$a + b = c$

8.1.4)

$a + b = c$

8.1.5)

8.1.6)

8.1.7)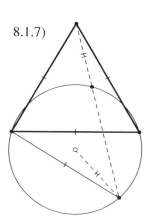

8.1.8)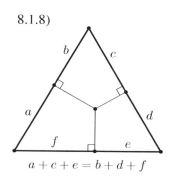

$a + c + e = b + d + f$

8.1.9)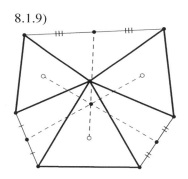

8.1.10)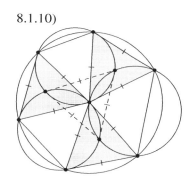

8.1.11)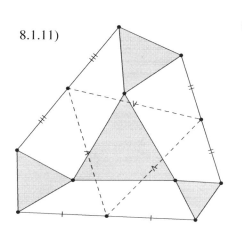

8.1.12)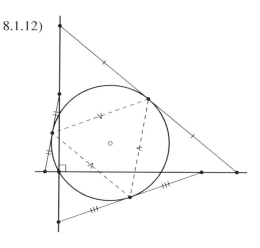

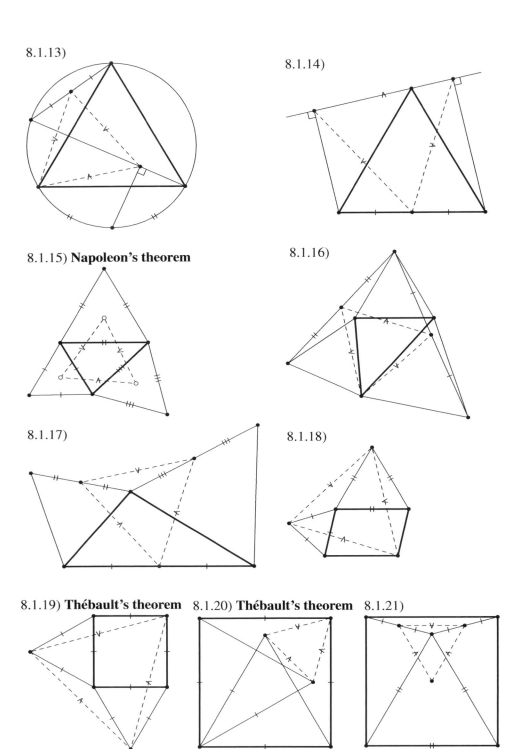

9 Appended polygons

9.1) **Napoleon point**

9.2)

9.3)

9.4)

9.5)

9.6)

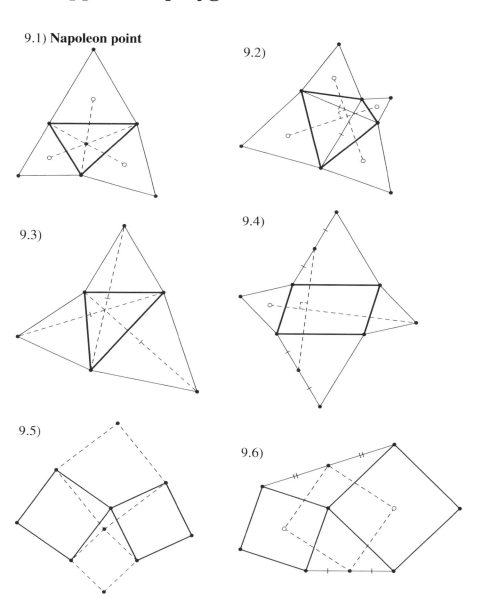

9.7)
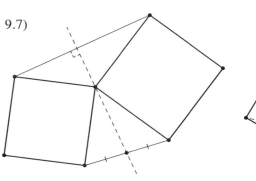

9.8)
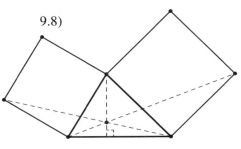

9.9)
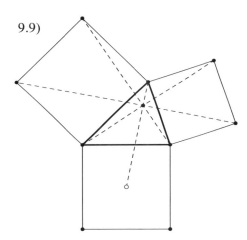

9.10)
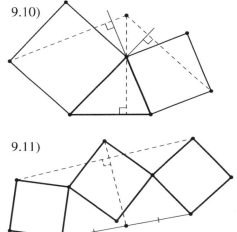

9.11)

9.12)
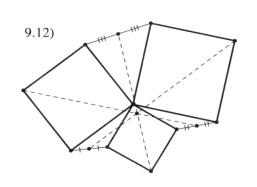

9.13)
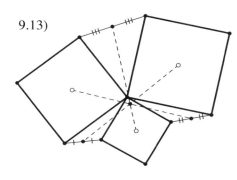

9.14) **Thébault's theorem**

9.15) **Van Aubel's theorem**

9.16)

9.17)

9.18)

9.19)

9.20)

9.21)

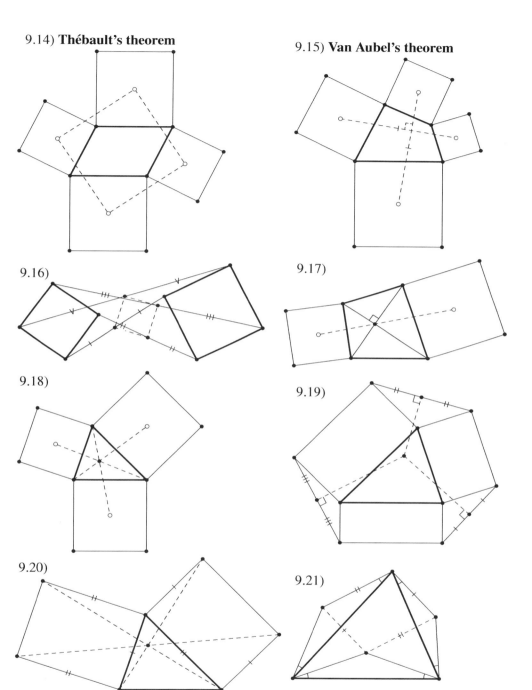

103

10 Chain theorems

10.1)

10.2)

10.3)

10.4)

10.5)

10.6)

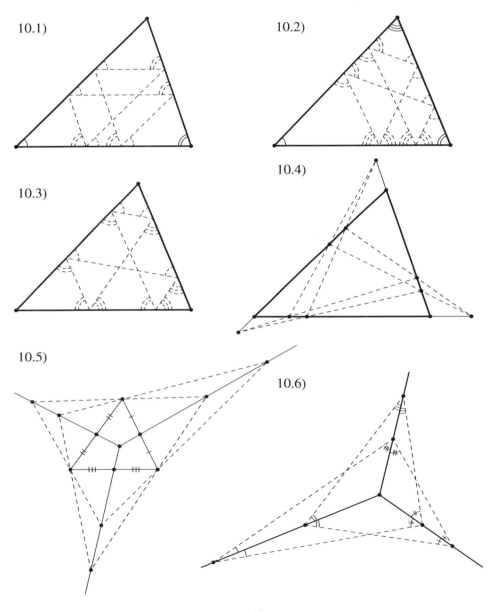

10.7)

10.8)

10.9)

10.10)

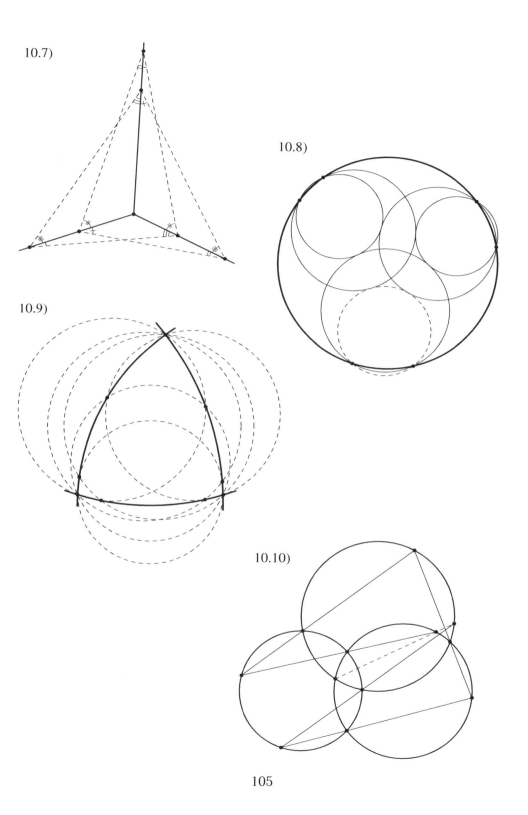

10.11)

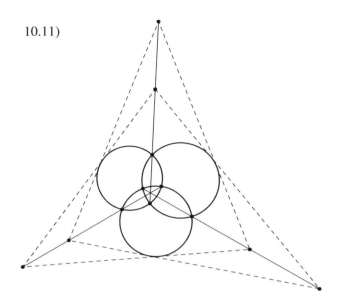

10.12)

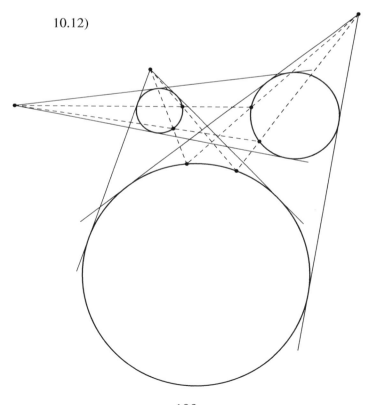

10.13)

10.14)

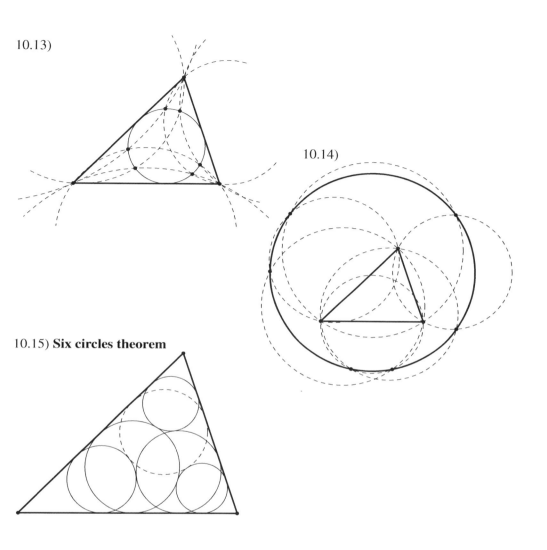

10.15) **Six circles theorem**

10.16) **Nine circles theorem**

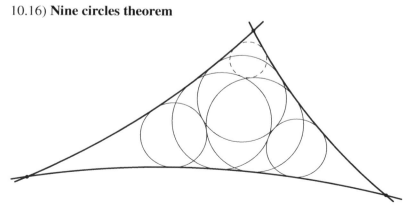

Poncelet's porism

10.17)

10.18)

10.19)

10.20)

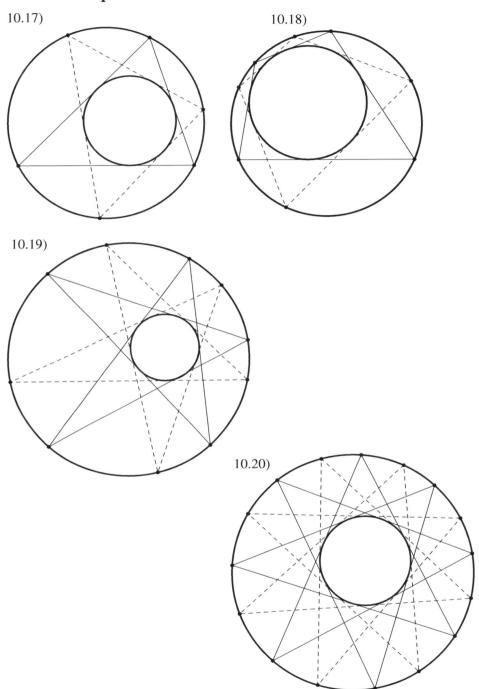

11 Remarkable properties of conics

11.1)

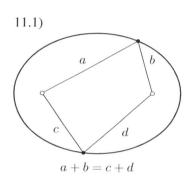

$a + b = c + d$

11.2)

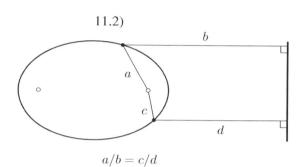

$a/b = c/d$

11.3) **Optical property of an ellipse**

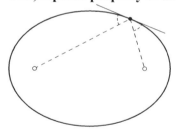

11.4)

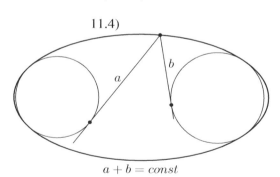

$a + b = const$

11.5) **Poncelet's theorem**

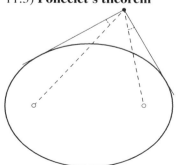

11.6)

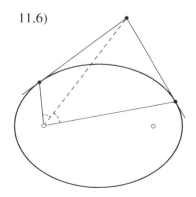

11.7)
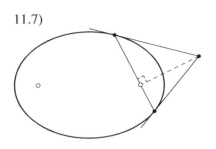

11.8)
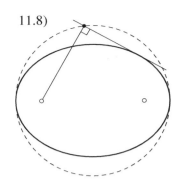

11.9)
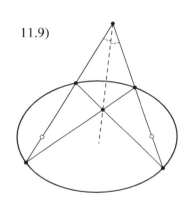

11.10)
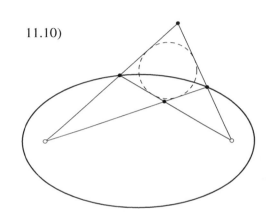

11.11)
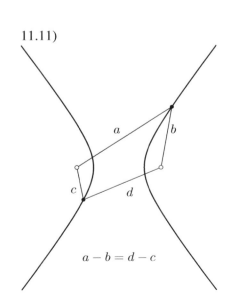

$a - b = d - c$

11.12) **Optical property of a hyperbola**
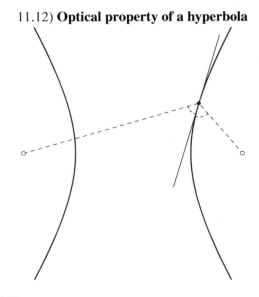

11.13)
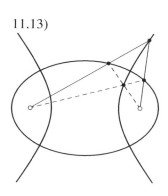

11.14)
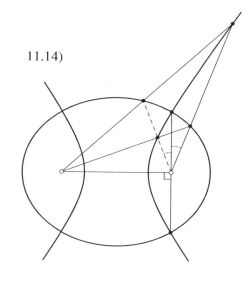

11.15)
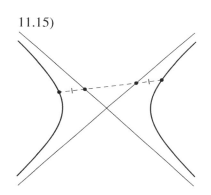

11.16) **Frégier's theorem**
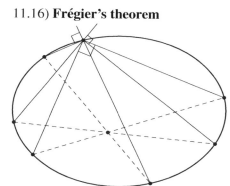

11.17)
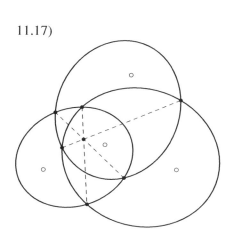

11.18) **Neville's theorem**
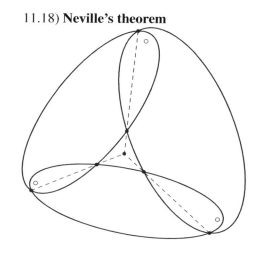

11.19) 11.20)

11.21) 11.22)

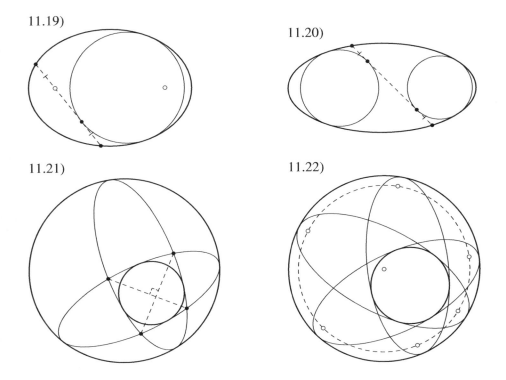

11.1 Projective properties of conics

Pascal's theorem

11.1.1) 11.1.2)

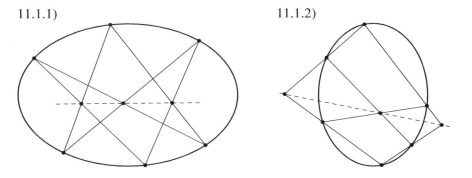

11.1.3)

11.1.4)

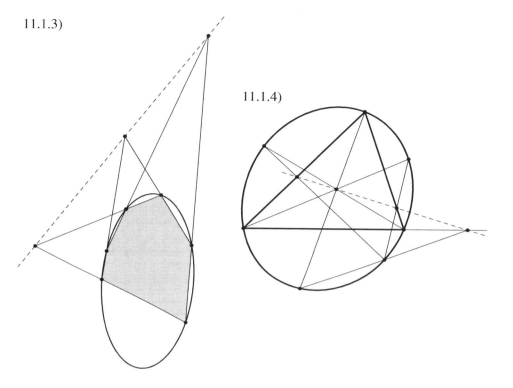

Brianchon's theorem

11.1.5)

11.1.6)

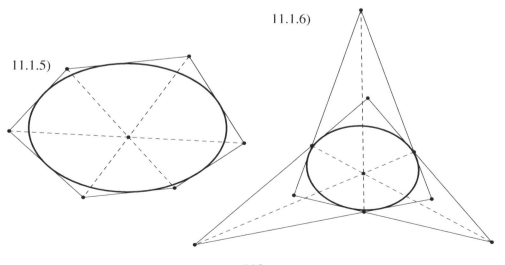

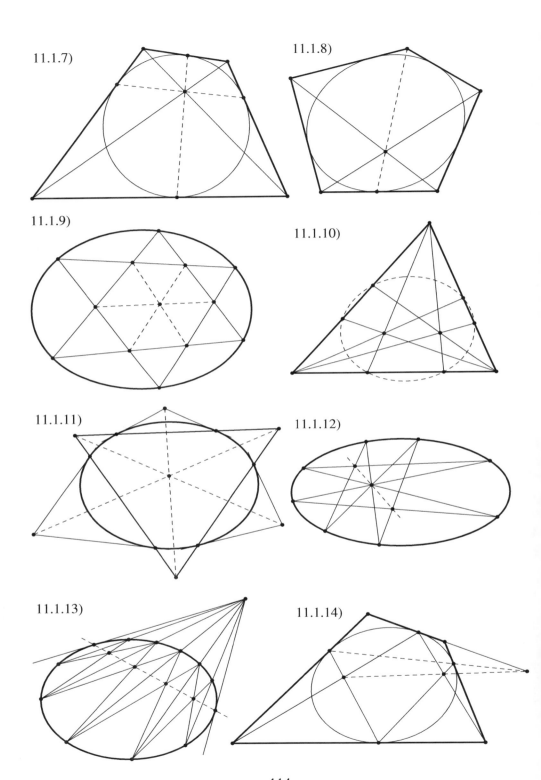

11.1.15)

11.1.17)

11.1.16) **Three conics theorem**

11.1.18)

11.1.19) **Dual three conics theorem**

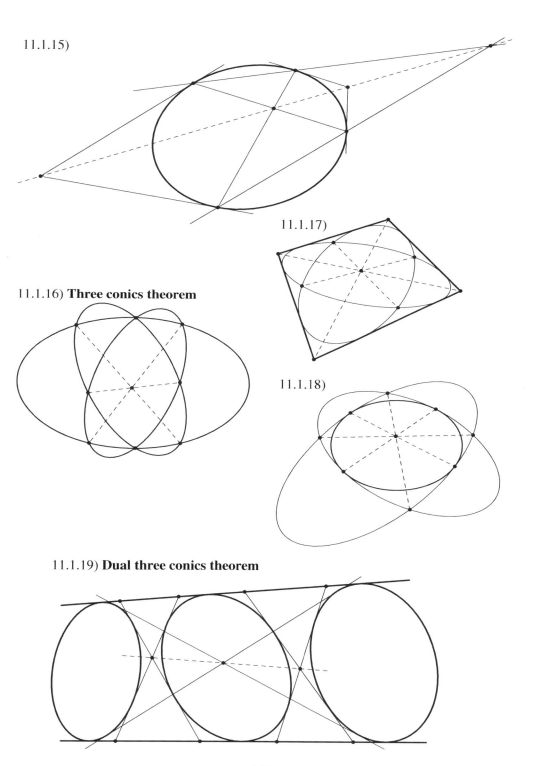

11.1.20) **Four conics theorem**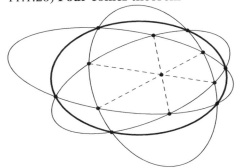

11.1.21)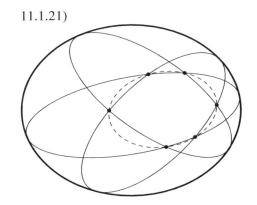

11.1.22)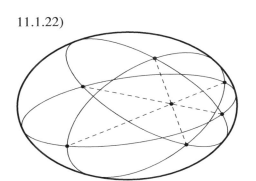

11.1.23)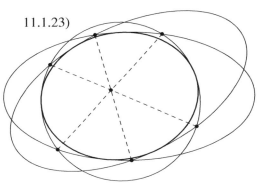

11.1.24)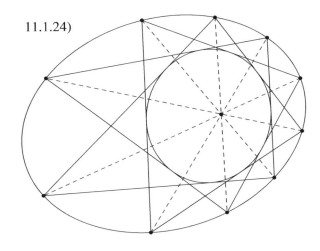

11.2 Conics intersecting a triangle

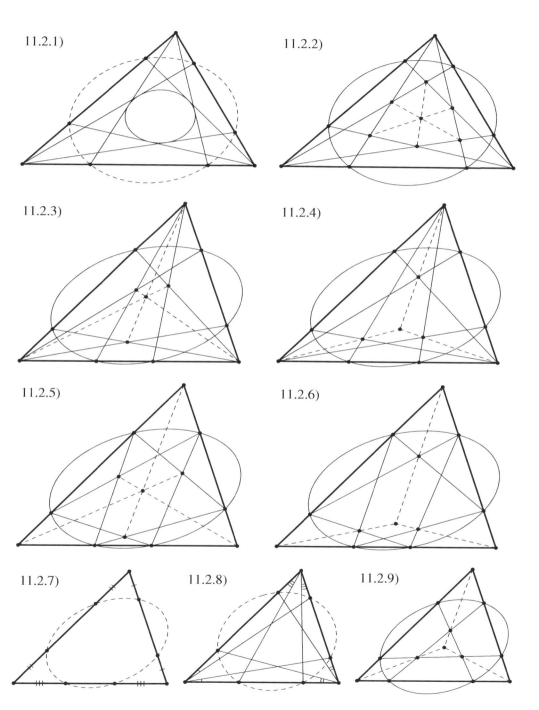

11.2.1)
11.2.2)
11.2.3)
11.2.4)
11.2.5)
11.2.6)
11.2.7)
11.2.8)
11.2.9)

11.3 Remarkable properties of the parabola

11.3.1)

11.3.2) **Optical property**

11.3.3)

11.3.4)

11.3.5)

11.3.6)

11.3.7)

11.3.8)

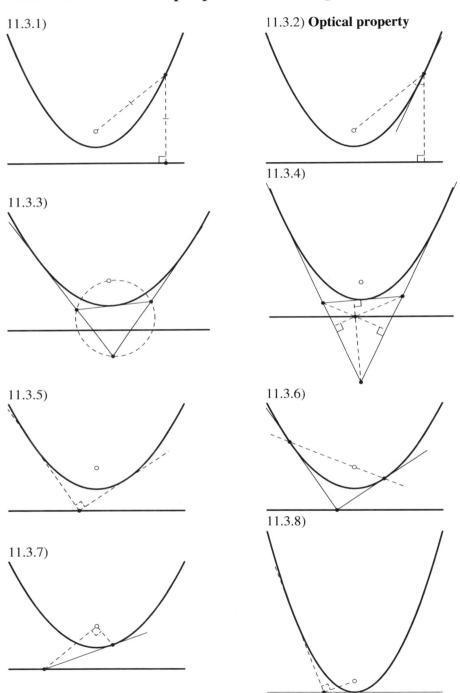

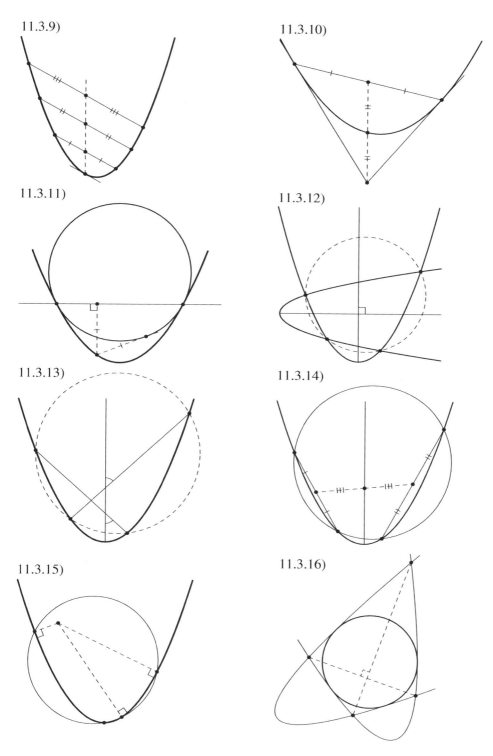

11.4 Remarkable properties of the rectangular hyperbola

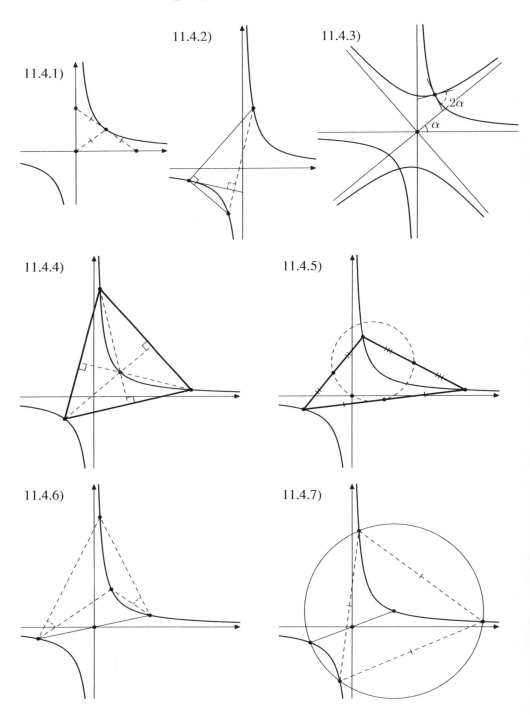

12 Remarkable curves

Lemniscate of Bernoulli

12.1)

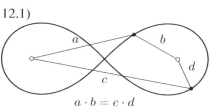

$a \cdot b = c \cdot d$

12.2)

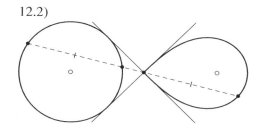

12.3)

Cissoid of Diocles

12.4)

12.5)

12.6)

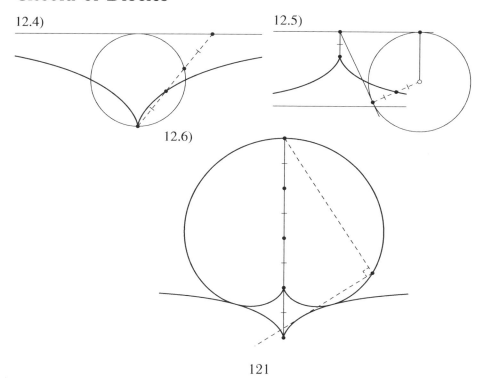

Cardioid

12.7)
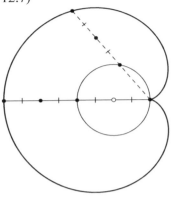

12.8)
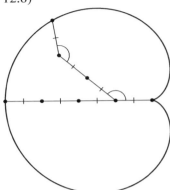

12.9)
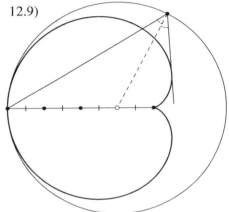

12.10)
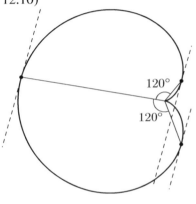

12.11)
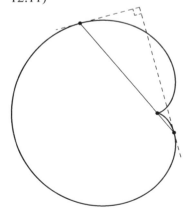

12.12)
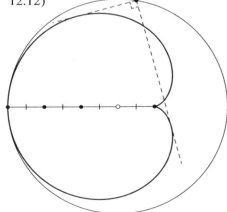

13 Comments

2.8) A. G. Myakishev, Fourth Geometrical Olympiad in Honour of I. F. Sharygin, 2008, Correspondence round, Problem 10.
2.32) Lemoine point is the center of the dashed circle.
3.9) Here it is shown that the Aubert line is perpendicular to the Gauss line.
4.1.2) A. A. Polyansky, All-Russian Mathematical Olympiad, 2007–2008, Final round, Grade 10, Problem 6.
4.1.3) L. A. Emelyanov, All-Russian Mathematical Olympiad, 2009–2010, Regional round, Grade 9, Problem 6.
4.1.9) V. V. Astakhov, All-Russian Mathematical Olympiad, 2006–2007, Final round, Grade 10, Problem 3.
4.1.17) See also 5.7.9.
4.1.19) A. V. Smirnov, Saint Petersburg Mathematical Olympiad, 2005, Round II, Grade 10, Problem 6.
4.1.20) D. V. Prokopenko, Fifth Geometrical Olympiad in Honour of I. F. Sharygin, 2009, Correspondence round, Problem 20.
4.1.21) A. A. Polyansky, All-Russian Mathematical Olympiad, 2010–2011, Final round, Grade 10, Problem 6.
4.1.23) M. Chirija, Romanian Masters 2006, District round, Grade 7, Problem 4.
4.2.6) United Kingdom, IMO Shortlist 1996.
4.3.6) R. Kozarev, Bulgarian National Olympiad, 1997, Fourth round, Problem 5.
4.3.8) Moscow Mathematical Olympiad, 1994, Grade 11, Problem 5.
4.3.10) D. Şerbǎnescu and V. Vornicu, International Mathematical Olympiad, 2004, Problem 1.
4.3.14) L. A. Emelyanov, All-Russian Mathematical Olympiad, 2009–2010, Final round, Grade 10, Problem 6.
4.3.17) I. I. Bogdanov, Sixth Geometrical Olympiad in Honour of I. F. Sharygin, 2010, Final round, Grade 8, Problem 4.
4.3.19) D. V. Prokopenko, All-Russian Mathematical Olympiad, 2009–2010, Regional round, Grade 10, Problem 3.
4.3.21) M. G. Sonkin, All-Russian Mathematical Olympiad, 1999–2000, District round, Grade 8, Problem 4.
4.4.6) A. A. Zaslavsky and F. K. Nilov, Fourth Geometrical Olympiad in Honour of I. F. Sharygin, 2008, Final round, Grade 8, Problem 4.
4.4.7) France, IMO Shortlist 1970.
4.5.5) USA, IMO Shortlist 1979.
4.5.12) Twenty First Tournament of Towns, 1999–2000, Fall round, Senior A-Level, Problem 4.
4.5.14) M. A. Kungozhin, All-Russian Mathematical Olympiad, 2010–2011, Final round, Grade 11, Problem 8.
4.5.15) Personal communication from L. A. Emelyanov.
4.5.16) Bulgaria, IMO Shortlist 1996.
4.5.20) F. L. Bakharev, Saint Petersburg Mathematical Olympiad, 2005, Round II, Grade 10, Problem 6.
4.5.22) Brazil, IMO Shortlist 2006. The bold line is parallel to the base of the triangle.

4.5.23) A. A. Polansky, All-Russian Mathematical Olympiad, 2006–2007, Final round, Grade 11, Problem 2. The bold line is parallel to the base of the triangle.
4.5.29) Special case of 6.3.3.
4.5.31) This construction using circles is not rare. See 10.15.
4.5.35) D. V. Shvetsov, Sixth Geometrical Olympiad in Honour of I. F. Sharygin, 2010, Correspondence round, Problem 8.
4.5.36) M. G. Sonkin, From the materials of the Summer Conference Tournament of Towns "Circles inscribed in circular segments and tangents", 1999.
4.5.37) M. G. Sonkin, All-Russian Mathematical Olympiad, 1998–1999, Final round, Grade 9, Problem 3.
4.5.38, 4.5.39) Based on Bulgarian problem from IMO Shortlist 2009.
4.5.40) D. Djukić and A. V. Smirnov, Saint Petersburg Mathematical Olympiad, 2005, Round II, Grade 9, Problem 6.
4.5.41) L. A. Emelyanov, Twenty Third Tournament of Towns, 2001–2002, Spring round, Senior A-Level, Problem 5.
4.5.43) V. A. Shmarov, All-Russian Mathematical Olympiad, 2007–2008, Final round, Grade 11, Problem 7.
4.6.2) A. I. Badzyan. All-Russian Mathematical Olympiad, 2004–2005, District round, Grade 9, Problem 4.
4.6.4) V. P. Filimonov, Moscow Mathematical Olympiad, 2008, Grade 11, Problem 4.
4.6.6) V. Yu. Protasov, Third Geometrical Olympiad in Honour of I. F. Sharygin, 2006, Correspondence round, Problem 15.
4.7.6) Generalization of 4.7.1.
4.7.8) Iranian National Mathematical Olympiad, 1999.
4.7.9) Iranian National Mathematical Olympiad, 1997, Fourth round, Problem 4.
4.7.18) Nguyen Van Linh, From forum www.artofproblemsolving.com, Theme: "A concyclic problem" at 27 May 2010.
4.7.16) Personal communication from K. V. Ivanov.
4.8.5) Personal communication from L. A. Emelyanov and T. L. Emelyanova.
4.8.7) Personal communication from F. F. Ivlev.
4.8.8) China, Team Selection Test, 2011.
4.8.9) L. A. Emelyanov, Journal "Matematicheskoe Prosveschenie", Tret'ya Seriya, N 7, 2003, Problem section, Problem 8.
4.8.13) A. V. Smirnov, Saint Petersburg Mathematical Olympiad, 2009, Round II, Grade 10, Problem 7.
4.8.15) G. B. Feldman, Seventh Geometrical Olympiad in Honour of I. F. Sharygin, 2011, Correspondence round, Problem 22.
4.8.16) L. A. Emelyanov and T. L. Emelyanova, All-Russian Mathematical Olympiad, 2010–2011, Final round, Grade 9, Problem 2.
4.8.20) A. V. Gribalko, All-Russian Mathematical Olympiad, 2007–2008, District round, Grade 10 Problem 2.
4.8.21) Special case of 4.8.23.
4.8.27) V. P. Filimonov, All-Russian Mathematical Olympiad, 2007–2008, Final round, Grade 9, Problem 7.
4.8.29) China, Team Selection Test, 2010.
4.8.30) T. L. Emelyanova, All-Russian Mathematical Olympiad, 2010–2011, Regional round, Grade 10, Problem 2.

4.8.32) A. V. Akopyan, All-Russian Mathematical Olympiad, 2007–2008, Grade 10, Problem 3.
4.8.33) D. Skrobot, All-Russian Mathematical Olympiad, 2007–2008, District round, Grade 10, Problem 8.
4.8.38) F. K. Nilov, Special case of problem from Geometrical Olympiad in Honour of I. F. Sharygin, 2008, Final round, Grade 10, Problem 7.
4.8.39) V. P. Filimonov, All-Russian Mathematical Olympiad, 2006–2007, Final round, Grade 9, Problem 6.
4.9.1) The obtained point is called *the isogonal conjugate with respect to the triangle*.
4.9.3) The obtained point is called *the isotomic conjugate with respect to the triangle*.
4.9.20) This point will be the isogonal conjugate with respect to the triangle. See 4.9.1.
4.9.26) A. A. Zaslavsky, Third Geometrical Olympiad in Honour of I. F. Sharygin, 2007, Final round, Grade 9, Problem 3.
4.10.4) D. V. Shvetsov, Sixth Geometrical Olympiad in Honour of I. F. Sharygin, 2010, Correspondence round, Problem 2.
4.10.6) A. V. Smirnov, Saint Petersburg Mathematical Olympiad, 2005, Round II, Grade 10, Problem 2.
4.11.2) D. V. Prokopenko, All-Russian Mathematical Olympiad, 2009–2010, Regional round, Grade 9, Problem 4.
4.11.9) S. L. Berlov, Saint Petersburg Mathematical Olympiad, 2007, Round II, Grade 9, Problem 2.
4.12.2) Generalization of Blanchet's theorem (see 4.12.1).
4.12.3) A. V. Smirnov, Saint Petersburg Mathematical Olympiad, 2004, Round II, Grade 9, Problem 6.
4.12.4) The bold line is parallel to the base of the triangle.
4.12.7) USSR, IMO Shortlist 1982.
5.1.1) M. A. Volchkevich, Eighteenth Tournament of Towns, 1996—1997, Spring round, Junior A-Level, Problem 5.
5.1.2) L. A. Emelyanov, All-Russian Mathematical Olympiad, 2000–2001, District round, Grade 9, Problem 3.
5.1.4) M. V. Smurov, Nineteenth Tournament of Towns, 1997–1998, Spring round, Junior A-Level, Problem 2.
5.1.5) V. Yu. Protasov, Second Geometrical Olympiad in Honour of I. F. Sharygin, 2006, Final round, Grade 8, Problem 3.
5.1.9) L. A. Emelyanov and T. L. Emelyanova, All-Russian Mathematical Olympiad, 2010–2011, Final round, Grade 11, Problem 2.
5.2.3) S. V. Markelov, Sixteenth Tournament of Towns, 1994–1995, Spring round, Senior A-Level, Problem 3.
5.2.5) A. A. Zaslavsky, First Geometrical Olympiad in Honour of I. F. Sharygin, 2005, Final round, Grade 10, Problem 3.
5.2.8) A. A. Zaslavsky, Third Geometrical Olympiad in Honour of I. F. Sharygin, 2007, Correspondence round, Problem 14.
5.2.10) A. V. Akopyan, Moscow Mathematical Olympiad, 2011, Problem 9.5.
5.2.2) The more general construction is illustrated in 5.4.16.
5.3.2) United Kingdom, IMO Shortlist 1979.
5.4.1–5.4.4) Special case of 5.4.5.
5.4.7) M. G. Sonkin, All-Russian Mathematical Olympiad, 1998–1999, Final round, Grade 11, Problem 3.
5.4.9) I. Wanshteyn.

5.4.10) A. A. Zaslavsky, Fourth Geometrical Olympiad in Honour of I. F. Sharygin, 2008, Correspondence round, Problem 10.
5.4.13) This construction is dual to the butterfly theorem. See 6.4.3 and 6.4.4.
5.5.2) F. V. Petrov, Saint Petersburg Mathematical Olympiad, 2006, Round II, Grade 11, Problem 3
5.5.3) W. Pompe, International Mathematical Olympiad, 2004, Problem 5.
5.5.8) M. I. Isaev, All-Russian Mathematical Olympiad, 2006–2007, District round, Grade 10, Problem 4.
5.5.9) P. A. Kozhevnikov, All-Russian Mathematical Olympiad, 2009–2010, Final round, Grade 11 Problem 3.
5.6.1, 5.6.2) I. F. Sharygin, International Mathematical Olympiad, 1985, Problem 5.
5.6.10) Personal communication from L. A. Emelyanov.
5.6.17) A. A. Zaslavsky, Twentieth Tournament of Towns, 1998–1999, Spring round, Senior A-Level, Problem 2.
5.7.4) Poland, IMO Shortlist 1996.
6.1.7) P. A. Kozhevnikov, International Mathematical Olympiad, 1999, Problem 5.
6.2.10) R. Gologan, Rumania, Team Selection Test, 2004.
6.5.9) V. B. Mokin. XIV The A. N. Kolmogorov Cup, 2010, Personal competition, Senior level, Problem 5.
6.6.3) A. A. Zaslavsky, Second Geometrical Olympiad in Honour of I. F. Sharygin, 2006, Final round, Grade 8, Problem 3.
6.7.9) Twenty Fourth Tournament of Towns, 2002–2003, Spring round, Senior A-Level, Problem 4
6.8.11) Constructions satisfying the condition of the figure are not rare (see 10.8).
6.9.1) I. F. Sharygin, International Mathematical Olympiad, 1983, Problem 2.
6.9.6) P. A. Kozhevnikov, Ninteenth Tournament of Towns, 1997–1998, Fall round, Junior A-Level Problem 4.
6.9.9) M. A. Volchkevich, Seventeenth Tournament of Towns, 1995–1996, Spring round, Junior A-Level, Problem 2.
6.10.1) M. G. Sonkin, All-Russian Mathematical Olympiad, Regional round, 1994–1995, Grade 9, Problem 6.
6.10.3) A. A. Zaslavsky, P. A. Kozhevnikov, Moscow Mathematical Olympiad, 1999, Grade 10, Problem 2.
6.10.4) P. A. Kozhevnikov, All-Russian Mathematical Olympiad, 1997–1998, District round, Grade 9, Problem 2.
6.10.7) Dinu Şerbănesku, Romanian, Team Selection Test for Balkanian Mathematical Olympiad.
6.10.8) France, IMO Shortlist, 2002.
6.10.10) Twenty Fifth Tournament of Towns, 2003–2004, Spring round, Junior A-Level, Problem 4
6.10.12) China, Team Selection Test, 2009.
6.10.13) I. Nagel, Fifteen Tournament of Towns, 1993–1994, Spring round, Junior A-Level, Problem 2. See also 4.3.15.
6.10.20) V. Yu. Protasov, Third Geometrical Olympiad in Honour of I. F. Sharygin, 2007, Final round, Grade 10, Problem 6.
6.10.21) USA, IMO Longlist 1984.
6.10.23) Personal communication from E. A. Avksent'ev. This construction is a very simple way to construct the Apollonian circle.
7.2) The obtained line is called *the trilinear polar with respect to the triangle*.
8.1.1) Here the points are reflections of the given point with respect to the sides of the triangle.
8.1.4) Bulgaria, IMO Longlist 1966. See also 6.1.10.

8.1.11) Hungary, IMO Longlist 1971.
8.1.12) E. Przhevalsky, Sixteenth Tournament of Towns, 1994–1995, Fall round, Junior A-Level, Problem 3.
8.1.13) I. Nagel, Twelfth Tournament of Towns, 1990–1991, Fall round, Senior A-Level, Problem 2.
9.1) If we construct our triangles in the direction of the interior, then we similarly obtain a point called *the second Napoleon point*.
9.2) Columbia, IMO Shortlist 1983.
9.3) See also 2.9 and 2.10.
9.9) Hungary–Israel Binational Olympiad, 1997, Second Day, Problem 2.
9.16) Belgium, IMO Longlist 1970. See alsos 4.12.9.
9.20) Twenty Seventh Tournament of Towns, 2005–2006, Spring round, Junior A-Level, Problem 3.
9.21) Belgium, IMO Shortlist 1983.
10.4) This construction is equivalent to that of the Pappus theorem.
10.7) Personal communication from F. V. Petrov.
10.13) Personal communication from V. A. Shmarov.
10.14) Generalization of previous result.
11.4) Personal communication from F. K. Nilov.
11.9) Personal communication from V. B. Mokin.
11.10) Personal communication from P. A. Kozhevnikov.
11.19, 11.20) Personal communication from F. K. Nilov.
11.21) Personal communication from F. K. Nilov.
11.22) Personal communication from F. K. Nilov.
11.1.13) This line is called *the polar line of a point with respect to the conic*.
11.1.24) The same statement holds for any inscribed circumscribed polygon with an even number of sides.
11.2.1, 11.2.2) Personal communication from A. A. Zaslavsky.
11.3.10) K. A. Sukhov, Saint Petersburg Mathematical Olympiad, 2005, Team Selection Test for All-Russian Mathematical Olympiad, Grade 10, Problem 1.
11.3.11) Personal communication from F. K. Nilov.
11.3.16) Personal communication from F. K. Nilov.

Constructions the author discovered while working on this book: 4.5.24, 4.5.25, 4.5.28, 4.5.31, 4.5.42, 5.4.17, 6.8.5, 6.8.6, 6.8.12, 10.8, 10.9, 10.11, 10.12, 10.14, 11.4.2.

Bibliography

1. N. Kh. Agakhanov, I. I. Bogdanov, P. A. Kozhevnikov, O. K. Podlipsky and D. A. Tereshin. *All-russian mathematical olympiad 1993–2006*. M.:MCCME, 2007.

2. A. V. Akopyan and A. A. Zaslavsky. *Geometry of conics*, volume 26 of *Mathematical World*. American Mathematical Society, Providence, RI, 2007.

3. M. Berger. *Geometry*. Springer Verl., 1987.

4. D. Djukić, V. Janković, I. Matić and N. Petrović. *The IMO compendium*. Springer, 2006.

5. D. Efremov. *New Geometry of the Triangle*. Odessa, 1902.

6. C. J. A. Evelyn, G. B. Money-Coutts and J. A. Tyrrell. *The seven circles theorem and other new theorems*. Stacey International Publishers, 1974.

7. R. A. Johnson. Advanced Euclidean Geometry, 2007.

8. V. V. Prasolov. *Problems in Plane Geometry*. M.:MCCME, 2006.

9. I. F. Sharygin. *Problems in Plane Geometry*. Mir Publishers, Moscow, 1988.

10. H. Walser. *99 Points of Intersection: Examples-Pictures-Proofs*. The Mathematical Association of America, 2006.

Made in the USA
Lexington, KY
18 October 2016